my **revision** notes

TOMORROW'S GEOGRAPHY

for **Edexcel** GCSE Specification A

Steph Warren

HODDER
EDUCATION
AN HACHETTE UK COMPANY

The publishers would like to thank the following for permission to reproduce copyright material:

Photo credits:

p2 © Steph Warren; **p3** Imagery copyright Getmapping plc/supplied by Skyscan.co.uk; **p13** © Steph Warren; **p14** © Crown copyright 2013. Ordnance Survey licence number 100036470; **p27** *m* & *b* © Steph Warren; **p32** *b* © Aleksey Bakaleev – Fotolia; **p33** © Steph Warren; **p44** © David McNew/Getty Images News; **p54** © Steph Warren; **p65** © Copyright Sasi Group (University of Sheffield) and Mark Newman (University of Michigan) (www.worldmapper.org); **p66** *(all)* Wikimedia Commons; **p67** Wikimedia Commons; **p68** © www.CartoonStock.com; **p115** *(all)* © Steph Warren; **p119** *(all)* © Steph Warren

Although every effort has been made to ensure that website addresses are correct at time of going to press, Hodder Education cannot be held responsible for the content of any website mentioned in this book. It is sometimes possible to find a relocated web page by typing in the address of the home page for a website in the URL window of your browser.

Hachette UK's policy is to use papers that are natural, renewable and recyclable products and made from wood grown in sustainable forests. The logging and manufacturing processes are expected to conform to the environmental regulations of the country of origin.

Orders: please contact Bookpoint Ltd, 130 Milton Park, Abingdon, Oxon OX14 4SB. Telephone: +44 (0)1235 827720. Fax: +44 (0)1235 400454. Lines are open 9.00a.m.–5.00p.m., Monday to Saturday, with a 24-hour message answering service. Visit our website at www.hoddereducation.co.uk.

© Steph Warren 2013

First published in 2013 by
Hodder Education,
an Hachette UK company
338 Euston Road
London NW1 3BH

Impression number 10 9 8 7 6 5 4 3
Year 2017 2016 2015 2014

Typeset in Cronos Pro Light 12pts by Datapage (India) Pvt. Ltd.
Cover image: © Ingram Publishing Limited; © Kadmy-fotolia; © DWP-fotolia; © dyvar-fotolia; © Geoff Pickering-fotolia
Artwork by Barking Dog Art, Datapage (India) Pvt. Ltd., Gray Publishing, Pantek Arts and Countryside Illustrations
Printed and bound in India

A catalogue record for this title is available from the British Library
ISBN 978 1 444 193879

Get the most from this book

This book will help you revise for the Edexcel A GCSE Geography specification. You can use the revision planner on pages iv–v to plan your revision, topic by topic. Tick each box when you have:

1 revised and understood a topic
2 answered the exam practice questions
3 checked your answers online.

You can also keep track of your revision by ticking off each topic heading throughout the book.

Tick to track your progress

Unit 1 Geographical Skills and Challenges

	Revised	Tested	Gone online
Chapter 1 Basic Skills			
1 Basic skills	☐	☐	☐
Chapter 2 Cartographic Skills			
4 Atlas maps	☐	☐	☐
6 Sketch maps	☐	☐	☐
7 Ordnance Survey (OS) maps	☐	☐	☐
Chapter 3 Graphical Skills			
15 Graphical skills	☐	☐	☐
Chapter 4 Geographical Enquiry and ICT Skills			
21 Geographical enquiry and ICT skills	☐	☐	☐
Chapter 5 Geographical Information System (GIS)			

Key terms

Key terms are highlighted in the text, with an explanation nearby.

Exam tips

Throughout the book there are exam tips that explain how you can boost your final grade.

Go online

Go online to check your answers and try out the quick quizzes at **www.therevisionbutton.co.uk/ myrevisionnotes.**

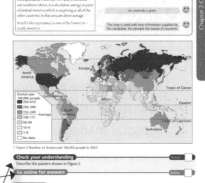

Check your understanding

Use these questions at the end of each section to make sure that you have understood every topic.

Exam practice

Sample exam questions are provided for each topic. Use them to consolidate your revision and practise your exam skills.

Contents and revision planner

Unit 1 Geographical Skills and Challenges

Unit 2 The Natural Environment

Introduction

Revision technique

Revise actively

There is a large amount of factual detail that you have to remember for your Geography exam, and if you only sit and read through your work, you may not be able to remember it. **Be active!** Here are some activities that might help your memory:

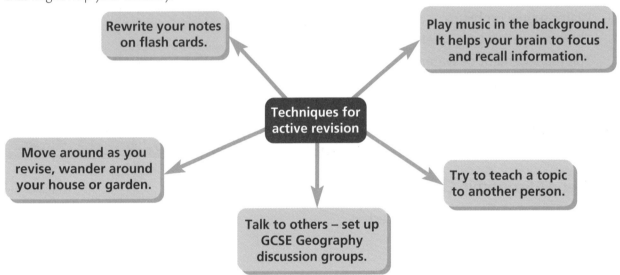

Rewrite your notes on flash cards.

Play music in the background. It helps your brain to focus and recall information.

Techniques for active revision

Move around as you revise, wander around your house or garden.

Try to teach a topic to another person.

Talk to others – set up GCSE Geography discussion groups.

Revision tips

- **Switch off the internet.** This stops you being distracted by social media sites and other websites.

- **Find your special place.** Allocate a room in your house as a working space. Your bedroom is probably not the best place!

- **Do short bursts of revision and reward yourself.** For example, do 20 minutes of revision for a reward of 10 minutes on your favourite social media sites.

- Relate work to an **anagram** or draw a diagram to help you remember.

- Don't forget to **eat** plenty of fruit and **drink** plenty of water.

- Put information on **sticky notes** around your mirror. You will read them subconsciously as you clean your teeth.

- **Practice papers and mark schemes.** Use these for revision, in order to become familiar with the wording of questions and how you answered or should have answered them.

- Think about where your weaknesses are and concentrate on revising for these topics.

Exam technique

Exam technique is all about how you complete the exam once you are in the examination hall.

- You should always read the front of the paper. This can sometimes be done while you are waiting for the exam to start if the paper is facing upwards on your desk.

- When you are told to start the exam, do not waste time putting your name on the front. Instead, start immediately and then put your name on the paper at the end.

- Do not waste time looking around in the exam, keep yourself focused and concentrating at all times.

www.therevisionbutton.co.uk/myrevisionnotes

Unit 1 exam techniques

The paper has 54 marks and a time limit of 60 minutes. Therefore, you must try to keep to a mark a minute. There is no choice on the paper, all questions have to be answered.

Start with Section B. This is the knowledge part of the exam. It will help you to settle into the paper because you will be doing a lot of writing and recalling the information that you have learnt. This will allow you to complete the 13 mark question (9 Geography and 4 SPaG marks) towards the middle of your allowed time when you are still fresh rather than when you are rushing at the end.

Now turn back to the beginning of the paper and complete the skills questions. If you are having problems with a question, put a star in the margin and carry on to the next question. You must, of course, remember to go back and complete the starred questions at the end.

Always check the answers you have given for all questions.

Units 2 and 3 exam techniques

These papers have 69 marks and time limits of 75 minutes. Therefore, you must try to keep to a mark a minute.

Start with Section B as you will have learnt one of these questions. By doing this, you will complete the SPaG allocation question towards the start of the exam. You will also have completed the question with the most marks allocated to it.

Then turn back to Section A. You must answer all of the questions in this section. There are three topics for the questions in Section A. Each topic has 15 marks awarded to it and ends with a question worth 6 marks on the higher tier paper and 4 marks on the foundation tier paper. Some people prefer to do all of the 4 or 6 mark questions at the same time but it is probably more logical to work through each topic in turn.

Command words

Here is a list of common command words which may be used on the exam paper. It is a good idea to underline the command words and any other key words in the question.

Compare	Say in what way two or more things are alike, or different from each other.
Contrast	Say in what way two or more things are different from each other.
Define	You may be asked to define a term. The examiner will be expecting you to state the meaning of that term in a geographical framework.
Describe	This is a very common command word and requires you to give the main characteristics of something. Questions will often ask you to describe a photograph, a pattern on a graph or a map. You should write an accurate account of what you see.
Name, give, identify or state	These words require you to answer briefly and are usually only worth 1 mark. For example, 'give the grid reference for …' or 'name one type of sea defence'.
Use data in your answer	This is often used with the command word 'describe'. In this case you must use data (information) in your answer. For example, you could be asked to describe the population distribution on a map using data in your answer. You would use the figures provided on the map to make specific factual comments on each area.
Discuss	If you are asked to discuss something, you will be expected to bring forward the important points of the argument.
Estimate	In some cases you may be asked to estimate a distance which means you have to give an approximate value.
Explain	This is another very common command word. It is asking you to give reasons as to **why** something occurs.
Justify	You may be required to justify your answer to a previous question, possibly using a map or a photograph. In this instance you must state the case for your answer, in other words, you must give reasons for your answer.
Outline	If you are asked to 'outline' something, you will be expected to summarise its main points.
Suggest / give reasons for	This is similar to the command word 'explain' but sometimes there are varying reasons why something happens and there is not necessarily a right or wrong answer. The examiner will expect you to give more than one reason.
Annotate	This means give a descriptive comment and an explanation.
Label	This is a simple descriptive comment which identifies something.
Rank	This means to put the answers into the correct order; you may also be required to justify your order.

Mark schemes

The following are examples of the types of mark schemes that will be used to mark your responses to the examination papers. You may find it helpful to refer to them when looking at the exam practice questions that are in this book.

An example mark scheme for 9 mark higher tier questions

Marks	Commentary
0 marks	No acceptable response.
1–3 marks	One or two basic points are made which are descriptive. There is no explanation. There is no specific detail about a location. Basic use of geographical terms.
4–6 marks	Good, clear descriptive points are made. Explanation is present but not developed. There is use of specific points with a located example. The work is clearly communicated with limited use of geographical terms.
7–9 marks	There are excellent descriptive points which are clearly about a number of specific locations with in-depth explanation. The response is well communicated with good use of geographical terms.

An example mark scheme for 6 mark higher tier questions

Marks	Commentary
0 marks	No acceptable response.
1–2 marks	A descriptive point is made. There is no explanation. There is no specific detail about a location.
3–4 marks	Good, clear descriptive points are made. All parts of the question are addressed but not equally. Explanation is present but not developed. The work is clearly communicated with limited use of geographical terms.
5–6 marks	There are excellent descriptive points which are clearly about a specific location with in-depth explanation. The response is well communicated with good use of geographical terms.

An example mark scheme for 6 mark foundation tier questions

Marks	Commentary
0 marks	No acceptable response.
1–2 marks	Has an idea about what the question refers to but gives no descriptive points. There is no specific detail about a location.
3–4 marks	One clear descriptive point that relates directly to the question. There is no explanation. A case study might be named but it is unclear on detail. The work is clearly communicated with limited use of geographical terms.
5–6 marks	Some good descriptive points which are clearly about a specific location. There is some clear explanation. The response is well communicated with good use of geographical terms.

An example mark scheme for 4 mark foundation tier questions

One mark will be given for each relevant point. Further marks are available for development of the point or for specific detail of a case study or example if that is required by the question. There is a maximum of 3 marks for basic points without development. Development means either further explanation or specific information.

An example mark scheme for SPaG (spelling, punctuation and grammar)

It is important to remember that some questions have spelling, punctuation and grammar marks awarded to them. This means that the examiner will be looking carefully at your spelling, punctuation and grammar and the way that you use geographical terminology.

The questions that give marks for SPaG are:
Unit 1 – the last question on the paper
Unit 2 – the last part of the question on Section B (Wasteful World or Watery World)
Unit 3 – the last part of the question on Section B (Moving World or Tourist's World)

Marks	Commentary
0 marks	Errors hinder the meaning of the response. The candidate does not use spelling, punctuation or the rules of grammar within the context of the question.
1 mark	Spelling, punctuation and use of grammar are reasonably accurate in the context of the question. Errors do not hinder the meaning of the response. Where required, a limited range of specialist terms are used appropriately.
2–3 marks	Spelling, punctuation and use of grammar are considerably accurate in the context of the question. Where required, a good range of specialist terms are used.
4 marks	Spelling, punctuation and use of grammar are consistently accurate in the context of the question. Where required, a wide range of specialist terms are used with precision.

Chapter 1 Basic Skills

Basic skills

How to label and annotate diagrams, graphs, sketch maps

Revised

You could be asked to do this in any of the units. For example, in Unit 2, you might get a question such as 'Draw an annotated diagram showing the formation of a waterfall or stack'.

In the Unit 1 exam, you could be asked to:

- complete the **annotations** on a diagram
- **label** a graph to show its main features
- annotate a sketch.

Key terms

Annotation – a label with more detailed description or an explanatory point.

Label – a simple descriptive point.

Exam tip

Ensure you have practice looking at Ordnance Survey (OS) maps and photographs of the same area so that you can identify:

- features on the map that are not on the photograph
- features on the photograph that are not on the map.

Check your understanding

Tested

1 Name one feature that is on Figure 1 (on page 2) but is not on the map on page 14.

2 Name one feature that is on the map on page 14 but is not in Figure 1.

Go online for answers

Online

How to draw, label and annotate sketches

Revised

For the exam you will usually need to show that you are able to complete a sketch. It is unlikely that you would be asked to draw the whole sketch because of time constraints.

Check your understanding

Tested

Figure 2 is an incomplete sketch of Figure 1. Complete the sketch by adding the rest of the hills, Buttermere Lake, the field pattern and Buttermere Village.

Go online for answers

Online

Exam tip

Follow the simple rules below:

- Some of the most important lines, such as rivers, coastline and the outline of the hills, will have been drawn for you. You could be asked to complete them.
- You could then be asked to add certain features such as woodlands, settlements or roads.
- You may possibly then be asked to label and annotate your sketch.

↑ Figure 1 A photograph of Buttermere

↑ Figure 2 An incomplete sketch map of Buttermere

How to interpret aerial, oblique and satellite photographs

Photographs show features of the landscape that are not on OS maps, such as the crops being grown in the fields. You might be asked to interpret aerial, oblique or satellite photographs but do you know the difference?

● **Aerial photographs** are taken from directly above, like the view of a flying bird.

● **Oblique photographs** are taken at an angle, so the detail of buildings can be seen.

● **Satellite photographs** are images taken from space. They show patterns of features, such as lights in an urban area. However, they can also show much more detail, such as the cars on a street.

When a photograph is interpreted, it involves describing and explaining the physical and human geography which can be seen on the photograph. It is important that the writing is coherent and shows good literacy skills when expressing geographical points.

Key terms

Aerial photographs are taken from directly above.

Oblique photographs are taken from above but at an angle.

Satellite photographs are images taken from space.

Exam tip

In the exam, you might be asked if a photograph is an oblique, aerial or satellite image.

↑ Figure 3

Check your understanding

1 Is Figure 3 an aerial, oblique or satellite image?

2 A village can be seen on the photograph. Describe the shape of the village. You can refer to the information on the shape of settlements on page 11 to help you.

3 What services might be in the village which cannot be seen on the photograph, but would be marked on an OS map?

Go online for answers

Online

Exam practice

Describe the human landscape shown on the photograph in Figure 3. **(3 marks)**

Answers online

Online

Atlas maps

How to describe the distribution or pattern of physical or human features on an atlas map

Revised ☐

Atlases contain maps which show **physical patterns**, such as the height of the land, and **human patterns**, such as population density. Atlas maps can be at different scales, showing features in a country through to global features.

Exam tip

In an exam you will need to be able to describe patterns of human geography and patterns of physical geography and relate them to each other.

When describing a distribution or pattern on an atlas map:

- start with a **general statement** about where the features are located on the map
- then go into **greater detail**, such as mentioning the area of the country or any particular features, for example the name of the sea next to that area.

It is a good idea to point out any **anomalies**.

Key terms

Physical features – the natural landscape.

Human features – the landscape created by people.

Anomalies – points that are different to the general trend.

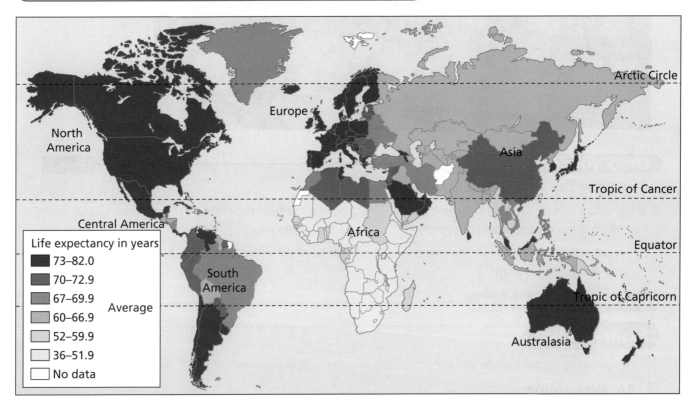

↑ **Figure 1** Life expectancy in 2003

Exam tip

Describe the distribution of life expectancy shown in Figure 1. (4 marks)

Life expectancy is high in North America, Australia, western Europe and parts of South America.

Life expectancy is low (36–51.9 years) in central and southern Africa. It is also below average in parts of Central America which is surprising as all of the other countries in that area are above average.

Brazil's life expectancy is one of the lowest in South America.

Information is given which describes the distribution. ☺

Data is given to support statements. ☺

An anomaly is given. ☺

The map is used with new information supplied by the candidate, for example the names of countries. ☺

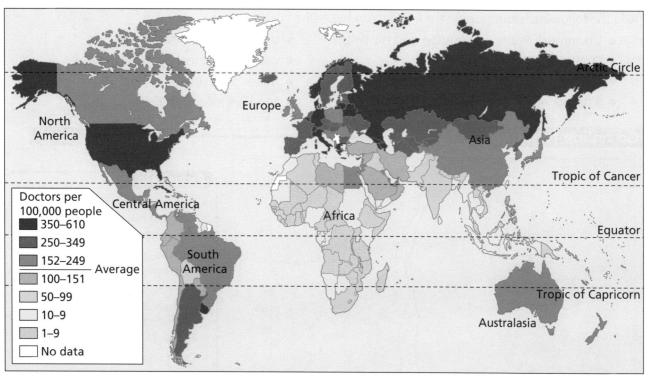

↑ Figure 2 Number of doctors per 100,000 people in 2004

Check your understanding
Tested ☐

Describe the pattern shown in Figure 2.

Go online for answers
Online ☐

Exam practice

Compare and contrast the information shown in Figure 1 and Figure 2. (4 marks)

Answers online

Online ☐

Sketch maps

How to draw, label, annotate, understand and interpret sketch maps

Revised

In the exam you may be asked to complete a sketch map. Remember, the examiner is not expecting a perfect replica of the map, but accuracy with the location of roads, woodlands or other features will be expected. If you are asked to complete a sketch map but not told what to include, just include the important features such as roads, railway lines, settlements and woods. Don't forget the title!

Check your understanding

Tested

Copy out and complete the sketch map in Figure 3.

Add the following features using the OS map on page 14:

- Crummock Water, Loweswater and the river connecting them
- B5289 and the rest of the secondary road network
- a mixed wood
- two camp and caravan sites
- national/regional cycle network
- three car parks
- 100-metre contour line.

Go online for answers

Online

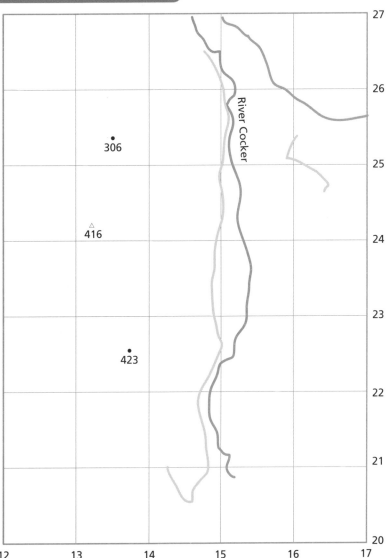

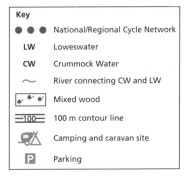

> **Exam tip**
>
> - In the exam it is best to use a pencil or black pen to complete the sketch map.
> - Use the symbols provided on the map key.
> - Use different types of lines such as dashed or dotted to show different features, for example roads.
> - Don't forget to provide a key!

↑ Figure 3 A sketch map of Lorton Vale, 2 cm = 1 km

Ordnance Survey (OS) maps

Map skills

Revised

The exam requires you to be competent in a lot of map skills, many of which may not appear on the exam paper, but you never know! See the list below:

- Symbols, four- and six-figure grid references
- Compass directions
- Straight line and winding distances
- Cross-sections
- Patterns of human and physical features
- The site, situation and shape of settlements
- Human activity from map evidence
- Use maps with photographs, sketches and written directions

Symbols, four- and six-figure grid references

Do I need to learn the symbols?

Many symbols provide a 'clue' to what they represent. For example a blue '**P**' is parking. Symbols will always be provided in a key with the 50,000 OS map, so you need to remember them. However, if you can learn some of them, it will save you time as you will not have to look them up.

Which grid references do I need to be able to use?

You are required to be able to use both four- and six-figure grid references.

When finding the grid reference 218405 (Figure 4), remember:

● Always go along and then up.

● The **first two** numbers (21) provide the line you require.

● The third number (8) is how far towards the next line you should go.

● The fourth and fifth numbers (40) provide the line up the side that you require.

● The sixth number (5) is how far towards the next line you should go.

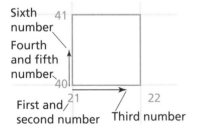

↑ **Figure 4 How to use a map's grid reference**

Compass directions

- These are normally examined on the OS map.
- You could be asked to orientate a photograph with the map and then provide a direction for a feature.
- Remember: north will be taken as the top of the map following the grid lines.

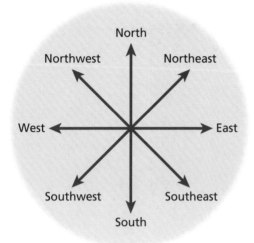

Straight line and winding distances

You may be asked to measure the distance between two points on a map along a road or a railway line. This could be a straight line or a winding distance.

- To measure a straight line, use a ruler or a piece of paper. Don't forget to use the scale line on the map to get the actual distance on the ground. You could also use a piece of string to measure distances. Lie the string along the route and measure against the scale line to get the distance required.
- To measure a winding distance, split the route into a number of straight sections. Then add up these distances and use the scale line to find out the distance on the ground.

> **Exam tip**
> It is useful to have a piece of string in the exam to measure distances.

> **Exam tip**
> Remember that the distance between grid lines on a 1:50,000 map (these are the only maps used in the exam paper) is 2 cm which is 1 km on the ground.

> **Exam tip**
> - You may also be asked to estimate the area of a feature, for example the area of Loweswater on the OS map extract on page 14.
> - Try to imagine whole grid squares. If you add the pieces of Loweswater not in 1221 in your mind to that grid square it would almost fill it.
> - Therefore the area of Loweswater is a square kilometre because that is the size of a grid square.

Check your understanding   Tested

Measure the length of the B5289 on the OS map extract (see page 14).

Go online for answers   Online

Cross-sections

Cross-sections show how relief varies along a chosen line on the map. A cross-section is drawn as a graph which shows distance on the x-axis (horizontal) and height on the y-axis (vertical). The scale must be chosen carefully so that a true representation of the area is produced.

> **Exam tip**
>
> In the exam it is unlikely that you would be asked to draw a cross-section but you may be asked to complete and/or annotate one. You should also be able to interpret cross-sections.
> - If the land is steep, the contours will be close together.
> - If the contours are far apart, the land is gently sloping.
> - An absence of contours indicates flat land.

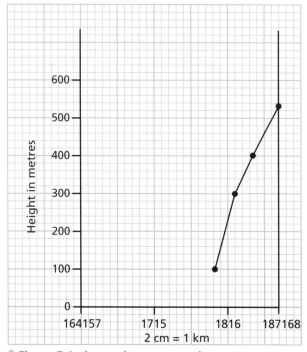

↑ **Figure 5 An incomplete cross-section**

Check your understanding

Figure 5 shows an incomplete cross-section between spot height 641 at grid reference 164157 to spot height 526 at grid reference 187168 (see the OS map on page 14).

1   Complete the cross-section.

2   Mark the following on your cross-section (label with an arrow and use a suitable key):
  - an area of deciduous woodland
  - an area of coniferous woodland
  - the B5289
  - Buttermere Lake.

3   What does your cross-section show about the relief and land use of the area?

Go online for answers

Patterns of physical and human features

There are a number of physical and human features which you could be asked to recognise on OS maps and to describe the patterns that they produce.

Physical features

Relief – this is the shape of the land

Ask yourself the following questions:

- Are there any contour lines?
- Are the contour lines close together or far apart?
- Have I included any actual figures from the map to illustrate the point I am making?

Vegetation

Ask yourself the following questions:

- What type of vegetation is on the map extract?
- What type of woodland is on the map extract?
- How much of the area is taken up with woodland?
- What other type of land use is there in the area?

Rivers and their valleys

Ask yourself the following questions:

- Are there any rivers or lakes in the area?
- Do the rivers make any particular patterns?
- Which direction is the river flowing in?
- Is there any human activity or interference with the rivers or lakes?

Exam tip

You might be asked to:

- describe a river and its valley
- compare two river valleys on a map extract.

Human features

Land use

Ask yourself the following questions:

- What type of settlement is in the area?
- What percentage of the area is taken up with settlement?

Communications

Ask yourself the following questions:

- Are the roads and railway lines in the river valleys?
- Are the railway lines close to the roads?

Check your understanding

Describe the pattern of woodland on the OS map on page 14.

Go online for answers

Exam tip

Remember the following when describing a distribution.

- Start with a **general statement** about where the features are located on the map.
- Then go into **greater detail**, such as mentioning grid references of where features are located.
- State **how much** of the map is taken up by the feature.
- If describing woodland, what type of woodland is on the map?
- Point out any **anomalies**.
- If you are asked to describe the distribution of settlement on a map, do not discuss individual settlement shapes.

The site, situation and shape of settlements

Revised

Site

You need to be able to describe the site and situation of settlements on a map. In order to describe or identify the site of a settlement there are certain human and physical features that you need to include.

An easy way to remember is using the acronym **SHAWL**. Do you know what the letters stand for?

Situation

The situation of a settlement is its position in relation to its surroundings. When you are describing the situation of a settlement on an OS map, you should describe the human and physical features around it. Try to remember the acronym **PARC**.

Shape

The shape of the settlement is the pattern that it makes. This refers to the way that the buildings are arranged. Settlement shape is usually concentrated on villages.

Site factors

Shelter from strong winds and storms
Height above sea level
Aspect – the way that the slope faces
Water supply
Land that the settlement is built on such as above the floodplain, fertile land, type of slope

Situation factors

Places
Accessibility
Relief
Communications

> **Exam tip**
> You could be asked to explain the site of a settlement using an OS map and a photograph.

Check your understanding — Tested

What is the shape of Low Lorton in grid square 1525 on the OS map on page 14?

Go online for answers — Online

Exam practice

Study the OS map on page 14. Describe the site of Buttermere. **(3 marks)**

Answers online — Online

> **Exam tip**
> You may be asked to compare the shape of two settlements on a map.

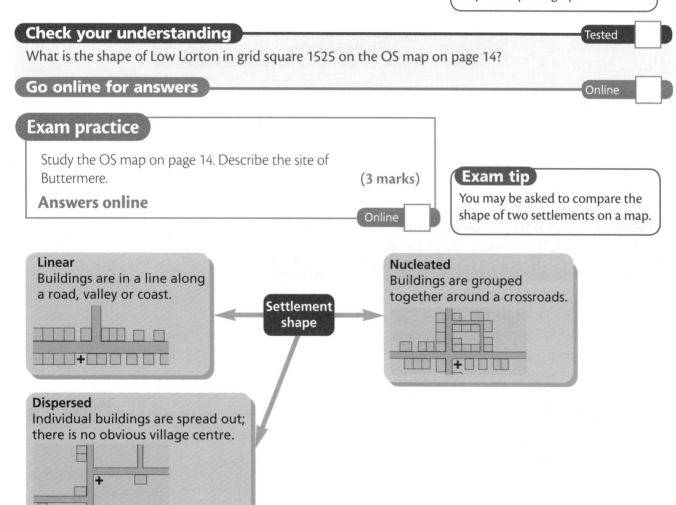

Linear
Buildings are in a line along a road, valley or coast.

Settlement shape

Nucleated
Buildings are grouped together around a crossroads.

Dispersed
Individual buildings are spread out; there is no obvious village centre.

Human activity from map evidence

Revised

Maps show a large amount of human activity. Most of it can be identified by looking at the key. In the exam, you could be asked to use information about human activity to produce sketch maps, or there may be questions relating to the distribution of different features.

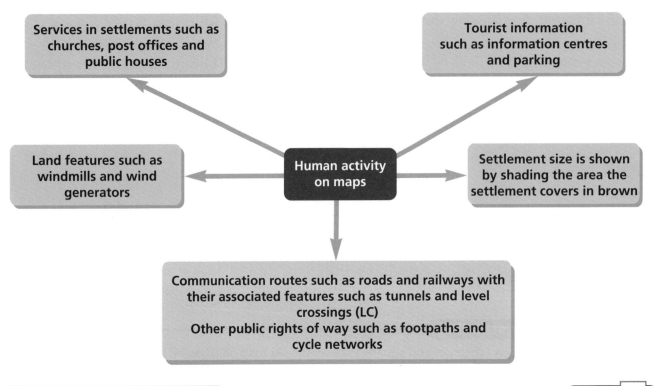

Services in settlements such as churches, post offices and public houses

Tourist information such as information centres and parking

Land features such as windmills and wind generators

Human activity on maps

Settlement size is shown by shading the area the settlement covers in brown

Communication routes such as roads and railways with their associated features such as tunnels and level crossings (LC)
Other public rights of way such as footpaths and cycle networks

Check your understanding
Tested

Using the OS map on page 14, give the six-figure grid references of five different types of tourist information.

Go online for answers
Online

Use maps with photographs, sketches and written directions
Revised

Many exam papers have questions which require you to be able to use the OS map with a photograph. This will require you to orientate the photograph with the map. The top of the map is always north – you must turn the photograph looking for important features until the photograph is pointing in the same direction as the map.

Check your understanding
Tested

Figure 6 was taken at grid reference 168148 on the OS map on page 14.

1 Name the two bodies of water (lakes) labelled A and B on the photograph.

2 What is the height of the hill at point C?

Go online for answers
Online

↑ Figure 6

Exam tip

The type of exam questions you could be asked are:

- recognition of certain features which have been identified by a letter on the photograph
- the direction the photograph was taken
- a comparison of features that can be seen on the map and not on the photograph, and vice versa
- the location of where the photograph was taken
- using the OS map to complete a sketch
- giving directions from one place to another, or following directions identifying features that you pass on the way
- identifying the best route between two places.

Exam practice

Using the OS map on page 14, provide the symbol and grid reference for   (6 marks)
a)   two tourist information features
b)   two general features.

Tourist information		General features	
Symbol	**Grid reference**	**Symbol**	**Grid reference**

Answers online

Online

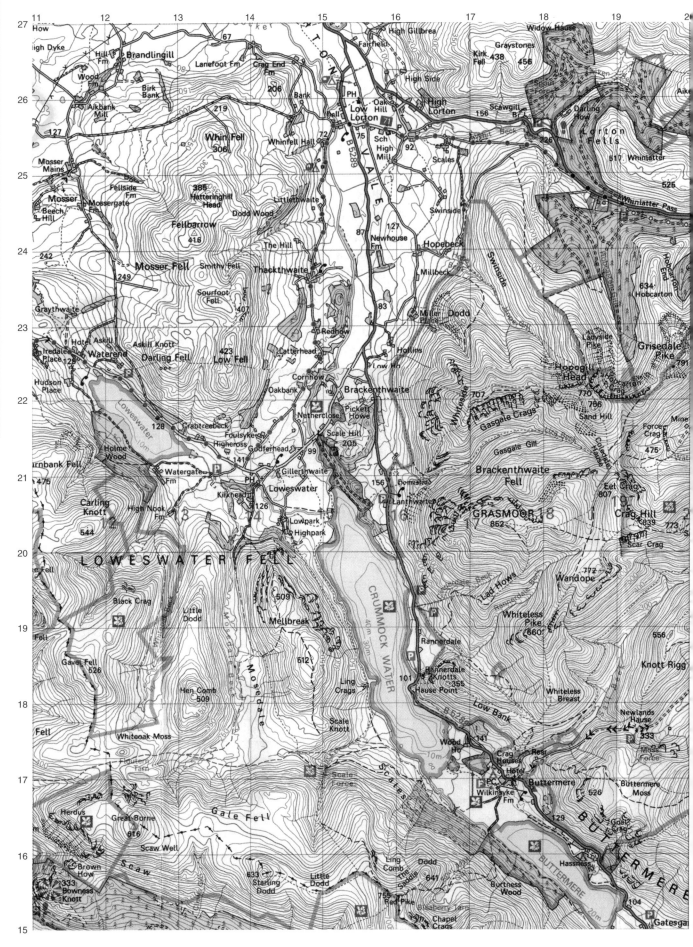

↑ **Figure 7 OS map of Loweswater**

Chapter 3 Graphical Skills

Graphical skills

The exam specification states that:

- you will need to be able to construct and complete a variety of graphs, charts and maps
- you will need to be able to interpret a variety of graphs, including those located on maps and topological diagrams.

The graphs listed below are the ones that you will be expected to know how to construct, complete and interpret. Some of the graphs are constructed in a similar way. So, how are your graphical techniques?

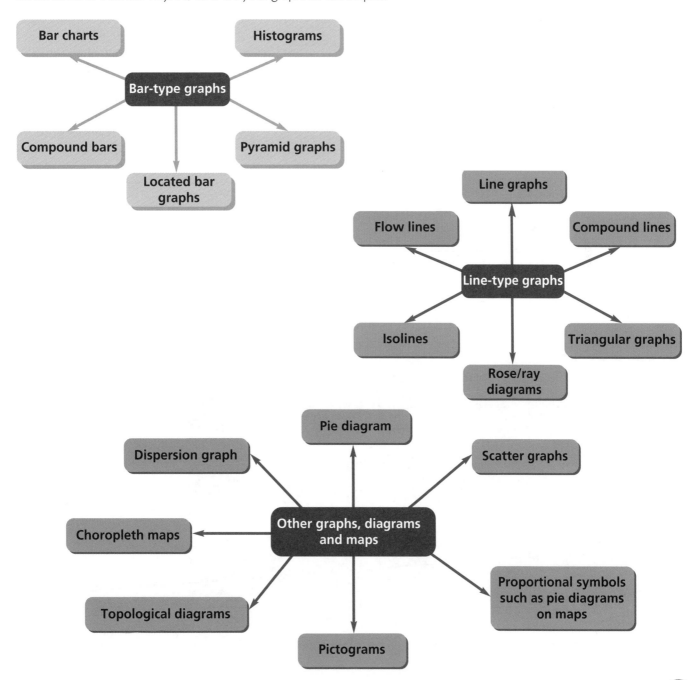

Bar-type graphs

Some of these will be very familiar to you such as **bar charts** and **histograms**. The difference between these is that a bar chart is used to display discrete (non-continuous) data and a histogram is used to display continuous data.

Bar charts and histograms can be drawn horizontally or vertically, the length of the column determining how many of the items are being displayed.

Other bar graphs are more difficult, such as compound bar graphs and bar graphs that are located on maps. A compound bar graph has a number of different pieces of information in each column. It could be just one column or have a number of columns, such as in Figure 1. This shows different employment sectors in a number of countries. By using a compound graph for this data the differences between the countries can be seen easily.

Key terms

Bar chart – a graph used to display discrete (non-continuous) data.

Histogram – a graph used to display continuous data.

Check your understanding

Describe the differences between the selected countries' sectors of employment.

Online

Exam tip

You could be asked to complete graphs or state the appropriateness of a graphical technique.

The compound bar graph shown in Figure 2 has been drawn with a gap between the areas that the data is being displayed for. This makes it easy to read the information. However, the patterns are easier to recognise if there is no gap between the bars.

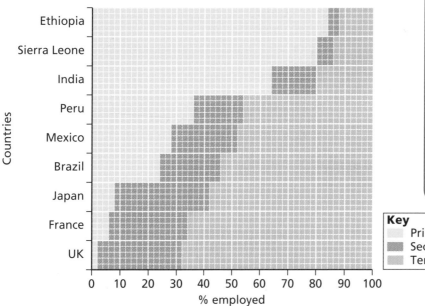

Key
Primary
Secondary
Tertiary

⬆ Figure 1 A compound or divided bar chart showing the sectors of industry in certain countries

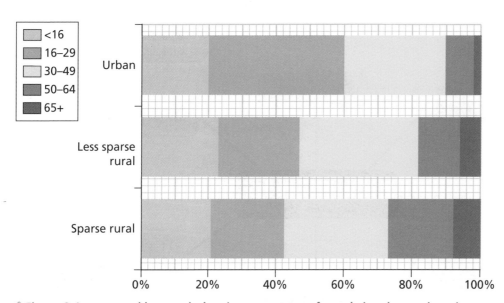

⬆ Figure 2 A compound bar graph showing age groups of people in urban and rural areas

Pyramid graph

The usual form of pyramid graphs is a population pyramid as shown in Figure 3. You could be asked to complete the graph or to interpret the graph. You would not be expected to give reasons for the changes that you state.

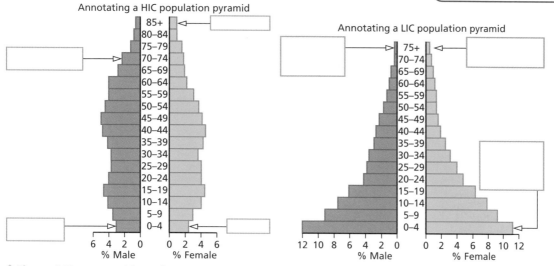

↑ **Figure 3 Two pyramid graphs**

Check your understanding

Draw two sketch pyramids which represent the pyramids in Figure 3.
Interpret the pyramids by labelling them with the following information.

- Women live longer than men.
- A narrow base shows a low birth rate.
- Wide bands at the top show a low death rate.
- There are fewer baby girls than boys.

- Few older people indicate a high death rate.
- A wide base shows a large number of children.
- There are more elderly women than men.

Tested

Go online for answers

Online

Bar graphs can also be located onto maps as in Figure 4. You will be expected to be able to construct bars on maps and to interpret them.

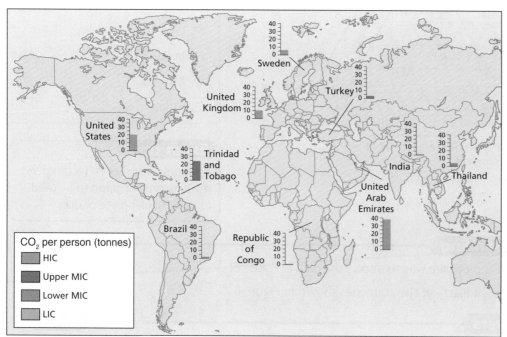

↑ **Figure 4 Carbon emissions for selected countries**

Line-type graphs

Line graphs are used to show data that is continuous. An example is Figure 5 which shows China's gross domestic product (GDP). The information is continuous because there is not a break in the years. Other types of line graphs are flow lines and isolines.

Flow lines are usually used to display some kind of movement such as pedestrian or traffic flows over time. **Isolines** are lines which join places which are equal, for example contours which join places of equal height and isovels which join places of equal velocity in a river.

Rose or ray diagrams tend to be used to show the direction of movement of groups of people. The length of the arrow would be the number of people and the arrow direction shows where the people come from.

Compound line graphs show continuous data for a number of variables. They can be some of the hardest graphs to interpret.

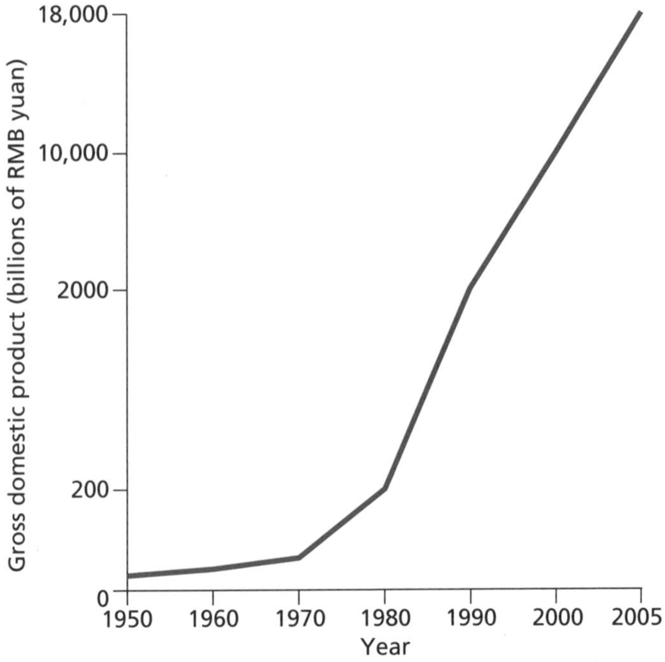

Triangular graphs show three variables on one graph, for example primary, secondary and tertiary industry for different countries. Figure 6 is an example of a triangular graph for sectors of employment.

↑ **Figure 5 China's GDP**

Key for Figure 6

1. Brazil
2. Bangladesh
3. Germany
4. Mali
5. Nepal
6. North Korea
7. Romania
8. Taiwan
9. UK
10. USA
11. India
12. Mexico
13. China

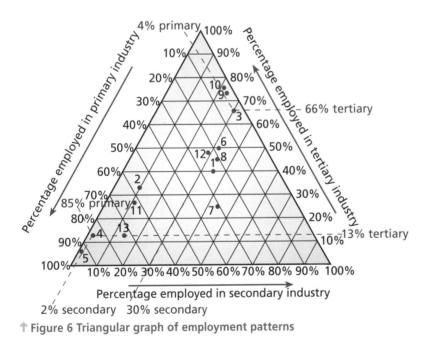

↑ **Figure 6 Triangular graph of employment patterns**

Exam tip

You could be asked to recommend the best type of graph to be used for a particular piece of data.

Check your understanding

Tested

1 Why is a line graph an appropriate way to display the data in Figure 5?

2 Describe the sectors of industry of the countries shown in Figure 6.

Go online for answers

Online

Other graphs, diagrams and maps

Pie diagrams and pictograms

A **pie diagram**, or divided circle, is a graphical technique for showing a quantity which can be divided into parts. Pie diagrams can be located on maps to show variations in the composition of a geographical phenomenon.

A **pictogram** is a way of portraying data using appropriate symbols or diagrams which are drawn to scale, as shown in Figure 7.

Another way of portraying data is **proportional symbols** on maps. These are usually circles but can also be squares or even symbols. The circles on Figure 8 are drawn to scale to portray the information on renewable energy.

Without a car	🚗	Key
With one car	🚗🚗🚗	🚗 =10%
With two or more cars	🚗🚗🚗🚗🚗🚗	

↑ **Figure 7 Car ownership**

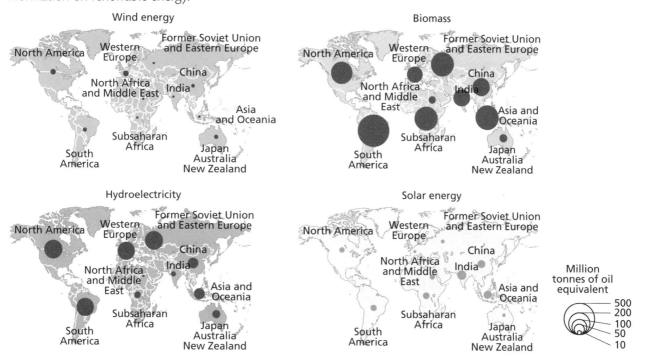

↑ **Figure 8 World potential renewable energy**

Scatter graphs

The other graphs that you need to be able to construct or interpret are **scatter graphs**. Scatter graphs show if there is any **correlation** between two sets of data. The correlation can be positive: as one set of data increases so does the other set of data; or negative: as one set increases, the other set decreases.

Exam practice

What are the advantages and disadvantages of using scatter graphs to display information? **(4 marks)**

Answers online

Online

Dispersion graphs, choropleth maps and topological diagrams

Dispersion graphs show the range of a set of data. **Choropleth maps** show data over an area. On **topological diagrams**, the position of the place remains the same but the distance and direction are not so important.

a

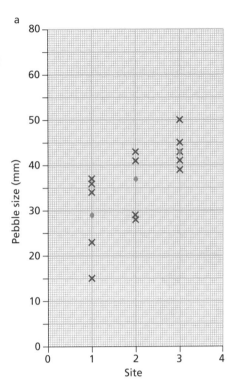

b

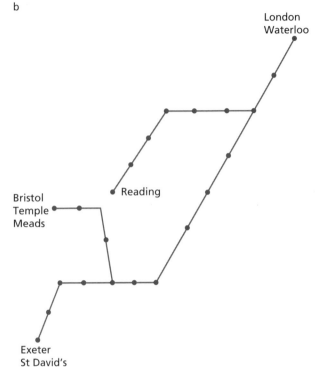

c

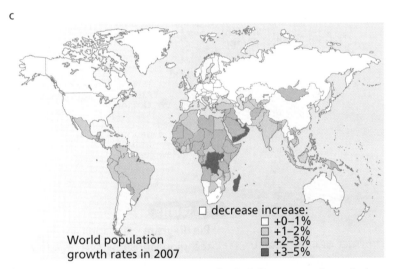

World population growth rates in 2007

decrease increase:
□ +0–1%
□ +1–2%
□ +2–3%
■ +3–5%

d

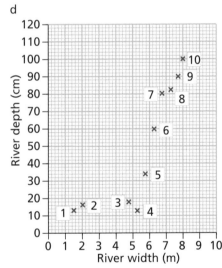

↑ Figure 9 A dispersion diagram, a topological diagram, a choropleth map and a scatter graph

Exam tip
You could be asked to recognise the type of graph being used.

Check your understanding   Tested

1 Complete the table below for Figure 9. Put a, b, c, d in the correct place in the table.

2 Justify your answer to question 1.

Type of display technique	
Topological	
Choropleth	
Scatter graph	
Dispersion	

Go online for answers   Online

Chapter 4 Geographical Enquiry and ICT Skills

Geographical enquiry and ICT skills

Geographical enquiry skills

Revised

There are a number of geographical enquiry skills which could be examined in the Unit 1 paper.

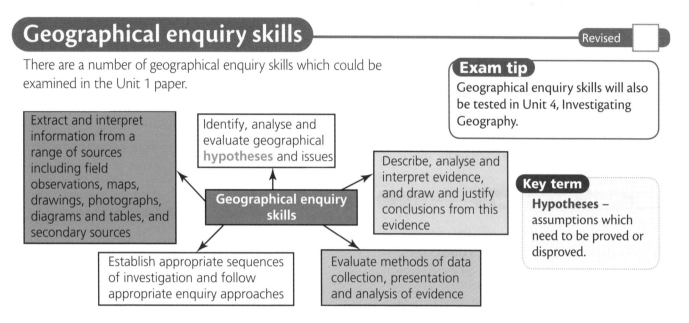

Extract and interpret information from a range of sources including field observations, maps, drawings, photographs, diagrams and tables, and secondary sources

Identify, analyse and evaluate geographical **hypotheses** and issues

Geographical enquiry skills

Describe, analyse and interpret evidence, and draw and justify conclusions from this evidence

Establish appropriate sequences of investigation and follow appropriate enquiry approaches

Evaluate methods of data collection, presentation and analysis of evidence

Exam tip

Geographical enquiry skills will also be tested in Unit 4, Investigating Geography.

Key term

Hypotheses – assumptions which need to be proved or disproved.

Some of these skills, for example, the ones mentioned below, are more likely to be examined than others.

Extract and interpret information from a range of sources including field observations, maps, drawings, photographs, diagrams and tables, and secondary sources

Check your understanding

Study the map of the River Thames drainage basin.

1. Which river flows through Luton?

2. The River Thames flows from point X to point Y. Name three cities that the River Thames flows through.

3. Which city does the River Thames flow through first?

Go online for answers

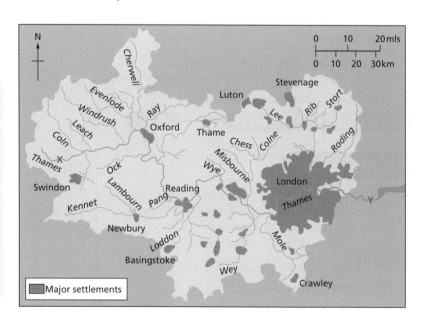

Describe, analyse and interpret evidence, and draw and justify conclusions from this evidence

Here, questions are asked about pedestrian flows recorded at different locations in Windsor. First, you are asked to describe the evidence; the second question asks for some interpretation of the evidence; the third question asks for concluding comments.

Study the pedestrian counts completed at different locations in Windsor, Berkshire.

1 Describe the information shown on the graph.

2 Suggest reasons for the patterns shown by the data.

3 What conclusions can be drawn from this information about the various locations in Windsor?

Go online for answers — Online

Evaluate methods of data collection, presentation and analysis of evidence

You could be asked to discuss the value of data collection and presentation techniques.

Check your understanding — Tested

Study the choropleth map of which county most people came from to Lulworth Cove in May 2007.

1 Suggest reasons why a choropleth map is a suitable technique to display this type of information.

2 Suggest one other technique that could have been used. Explain why it is a more appropriate technique than a choropleth map.

Go online for answers — Online

Exam practice

A choropleth has been used to display the information about people coming to Lulworth Cove. Suggest one other technique that could have been used. Give reasons for your choice.   **(3 marks)**

Answers online — Online

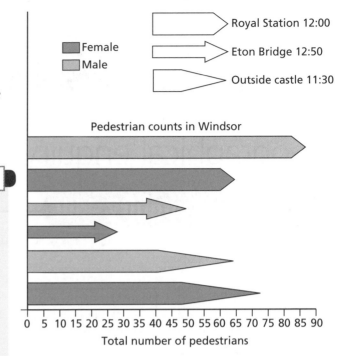

Royal Station 12:00
Eton Bridge 12:50
Outside castle 11:30

Female
Male

Pedestrian counts in Windsor

0 5 10 15 20 25 30 35 40 45 50 55 60 65 70 75 80 85 90

Total number of pedestrians

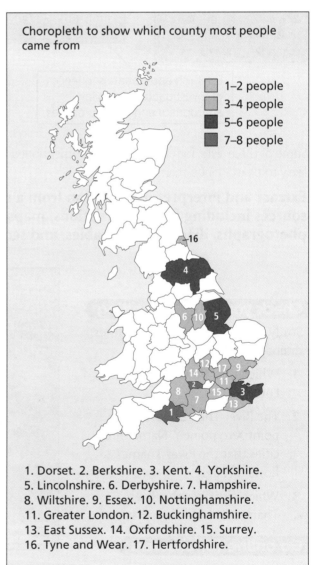

Choropleth to show which county most people came from

1–2 people
3–4 people
5–6 people
7–8 people

1. Dorset. 2. Berkshire. 3. Kent. 4. Yorkshire. 5. Lincolnshire. 6. Derbyshire. 7. Hampshire. 8. Wiltshire. 9. Essex. 10. Nottinghamshire. 11. Greater London. 12. Buckinghamshire. 13. East Sussex. 14. Oxfordshire. 15. Surrey. 16. Tyne and Wear. 17. Hertfordshire.

ICT skills

Revised

There are a number of ICT skills which are listed in the exam specification.

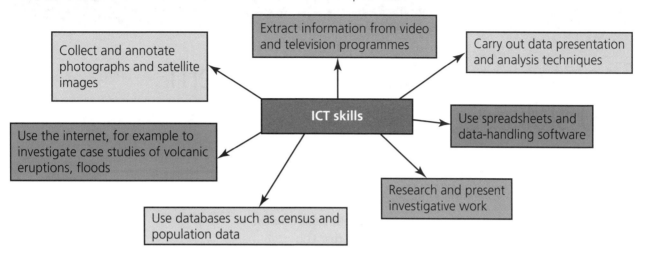

The skills could be examined in the following ways:

Research and present investigative work, for example, how you would carry out an investigation in the field.

Use spreadsheets and data-handling software, for example, to interpret a spreadsheet of data.

Collect and annotate photographs and satellite images, for example, annotate a photograph using information from a piece of stimulus material or perhaps the OS map.

Carry out data presentation and analysis techniques, for example, complete a graph such as a bar graph or a map such as a choropleth map, or carry out data analysis techniques such as producing a scatter graph.

Use databases such as census and population data, for example, interpret census data. This could be in the form of a database.

Chapter 5 Geographical Information System (GIS) Skills

Geographical information system (GIS) skills

Using GIS
Revised

There are a number of ways that GIS could be examined in Unit 1. You will not be asked to go to the computer room and capture information or use a web-mapping site, but there are lots of questions that you might be asked.

Check your understanding
Tested

1 What is GIS?

2 Who uses GIS?

3 What is layering?

4 How is GIS used?

5 How can GIS improve data presentation?

6 How do web-mapping sites such as Google work?

7 How have you used GIS?

Go online for answers
Online

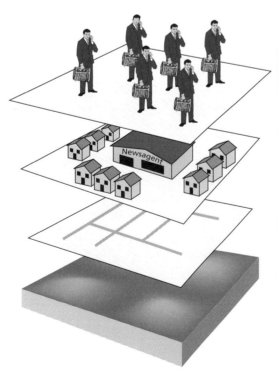

The people indicate where newspapers are delivered to each day.

This layer shows houses. The central building is a newsagent.

This layer shows the roads of the area.

Base map showing physical features.

Exam practice

What are the advantages and disadvantages of using GIS?   (4 marks)

Answers online
Online

Chapter 6 Challenges for the Planet

The causes, effects and responses to climate change

How has the world's climate changed since the last ice age?

Revised

The temperature over the last 10,000 years has increased by 5°C. However, there have been a number of fluctuations in this general trend, as is shown in Figure 1.

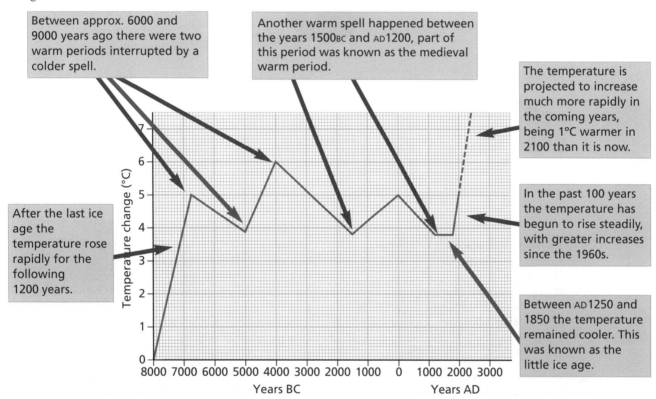

Between approx. 6000 and 9000 years ago there were two warm periods interrupted by a colder spell.

Another warm spell happened between the years 1500BC and AD1200, part of this period was known as the medieval warm period.

The temperature is projected to increase much more rapidly in the coming years, being 1°C warmer in 2100 than it is now.

After the last ice age the temperature rose rapidly for the following 1200 years.

In the past 100 years the temperature has begun to rise steadily, with greater increases since the 1960s.

Between AD1250 and 1850 the temperature remained cooler. This was known as the little ice age.

↑ **Figure 1 A graph of the world's temperature since 8000BC**

Key terms

Cause of climate change – a factor that makes climate change happen.
Effect of climate change – something that occurs as a result of climate change.
Response to climate change – the reaction that people are having to climate change.
Ice age – when the majority of the Earth's water is stored in glaciers as ice.

Exam tip

The exam could contain questions on how the climate has changed without providing any stimulus material so ensure that you can describe what has happened.

Exam practice

Describe how the world's climate has changed since the last ice age. Use Figure 1 to help you.   **(4 marks)**

Answers online

Online

Why has the world's climate changed since the last ice age?

Solar output is energy that comes from the Sun. Measurements that have been taken since the 1980s show that the amount of solar energy reaching the Earth has decreased by 0.1 per cent.

The activity of sunspots on the Sun's surface affects solar output. There was reduced sunspot activity between 1645 and 1715, which is thought to be one of the causes of the little ice age.

The Earth's orbit around the Sun varies from nearly circular to elliptical and back to circular again every 95,000 years. There are many conflicting views on what happens to temperatures as the Earth's orbit shape changes. Cold periods seem to have occurred when the Earth's orbit is circular and warmer periods when it is more elliptical.

Solar output

World's climate

Orbital geometry

Volcanic activity

Change in atmospheric gases

Volcanic eruptions release large amounts of sulphur dioxide and ash into the atmosphere. These act as a cloak and reduce the amount of solar (radiation) energy reaching the Earth's surface. In 1815, Mount Tambora erupted. The following year was unusually cold over much of the world with Europe having heavy snowfalls and frost throughout the summer.

There is a clear relationship between the amount of carbon dioxide (CO_2) in the atmosphere and temperature variations. Carbon dioxide is one of the most important gases responsible for the greenhouse effect. The greenhouse effect keeps heat within the Earth's atmosphere by absorbing longwave radiation. Without the greenhouse effect, the average global temperature of the Earth would be −18°C rather than the present 15°C.

↑ Figure 2 The factors that affect the world's climate

Key terms

Solar output – energy that comes from the Sun.

Orbital geometry – the movement of the Earth around the Sun and Moon.

Circular – in a circle.

Elliptical – in an egg shape.

Carbon dioxide – a gas released when fossil fuels are burnt.

Fossil fuels – fuels that are made from the remains of creatures and vegetation which have decomposed in the Earth's surface, for example coal and oil.

Exam tip

The causes of climate change in the section on this page have all been happening for thousands of years. Methane and the burning of **fossil fuels** (see page 27) are more recent causes of climate change.

The causes of current climate change on a local and global scale

Revised

Climate change has a number of different causes; these include the burning of fossil fuels and the increase of methane in the atmosphere.

Fossil fuels	Methane
When fossil fuels such as coal, oil and natural gas are burnt they produce carbon dioxide which contributes to the greenhouse effect causing the climate to change. • There has been an increase in the burning of fossil fuels to produce energy in countries such as China, where 75 per cent of energy is produced from coal. China is developing rapidly and using coal to fuel this development. • There has been an increase in global car ownership which means more oil is used, which in turn increases the amount of carbon dioxide in the atmosphere. • Drilling for oil also releases methane in the form of natural gas.	Methane makes up 20 per cent of the greenhouse gases in the atmosphere and is twenty times more potent than carbon dioxide. The amount of methane in the atmosphere has risen by 1.5 per cent a year for the past decade. But why? • There has been an increase in bacteria emissions from wetlands because of rising temperatures. • There has been an increase in the growing of rice because of the growing population in rice producing countries. Rice is grown in marshy conditions. • There has been an increase in the number of cattle for meat reflecting an increase in Western-style diets. Cattle produce methane as they ruminate their food.

↑ **Figure 3 Rice growing in a paddy field in Sri Lanka**

Check your understanding — Tested

Explain how the land uses shown in the photographs cause climate change.

Go online for answers — Online

Exam tip

Questions could be about general causes of climate change or concentrated on specific causes such as volcanic eruptions or fossil fuels.

↑ **Figure 4 Cows ruminating in a field in Devon**

The negative effects of climate change

Revised

Climate change has negative effects on the environment and people. These negative effects can be on both a local and a global scale. On a global scale there has been a change in crop yields, sea levels and glaciers:

Type of effect	Examples of where it is happening in the world
Crop yields If temperatures continue to rise it will have an effect on food production.	• **Tanzania** is experiencing longer periods of **drought**. Farmers will lose almost a third of their maize crop. • In **India** there will be a 50 per cent decrease in the amount of land available to grow wheat. This is due to hotter and drier weather. • In **Kenya**, droughts now happen every three years instead of every ten years. In 2006, Kenya suffered its worst drought for 80 years. Many farmers lost all of their cattle.
Sea levels and marine environments If the sea level continues to rise many areas of land will be flooded and some marine environments will be endangered.	• Due to rising sea levels, **Tuvalu** (a group of nine coral atolls in the Pacific Ocean) has started to evacuate its population to New Zealand, with 75 people moving away each year. Bangladesh suffers from coastal flooding. • Experts say if the sea level goes up by 1 m, **Bangladesh** will lose 17.5 per cent of its land. • Sea levels rising will threaten large areas of low lying coastal land including major world cities such as **London**. • Many of the coral reefs in the world, such as the Great Barrier Reef, are dying because of an increase in the temperature of the water caused by global warming.
Glaciers If temperatures continue to rise glaciers will melt at a faster rate.	• Glacier National Park, **Montana**, USA, was created in 1910. At this time there were 150 glaciers. Since then the number has decreased to 30. It is predicted that within 30 years most if not all of the park's glaciers will disappear. • In the **Antarctic** 90 per cent of the glaciers are retreating. • Melting ice in the Arctic could cause the **Gulf Stream** to be diverted further south. This will lead to colder temperatures in **western Europe**; temperatures would frequently be below 0°C in the winter with averages of 8–10°C in July, which is 10°C cooler than the average UK summer temperature.

Key terms

Drought – a long period without any precipitation (rainfall).

Gulf stream – a warm body of water that moves out of the Gulf of Mexico and crosses the Atlantic, warming up the coast of the UK.

Exam tip

Questions could be about the effects of climate change in general or could focus on one, such as the effect of climate change on food production.

Exam practice

Explain the negative effects of climate change. Use examples in your answer. **(4 marks)**

Answers online

Online

Check your understanding

Tested

Learn two specific examples from each of the types of effects in the table above.

Go online for answers

Online

The responses to climate change – from a global to a local scale

Revised

Global scale responses to climate change

Global agreements between nations

June 1992 – The Earth Summit, Rio de Janeiro

This was a meeting organised by the United Nations to discuss climate change. The result of the meeting was the first international environment treaty which aimed to stabilise greenhouse gas emissions.

December 1997 – Kyoto Conference

At this meeting the Kyoto Protocol was signed which came into force in February 2005. By 2008, 181 countries had signed the Kyoto Protocol. The agreement stated the following:

- Greenhouse gas emissions to be cut by 5.2 per cent compared to 1990 levels globally.
- Each country agreed to a national limit on emissions which ranged from 8 per cent for the EU, 7 per cent for the USA, 6 per cent for Japan and 0 per cent for Russia.

- It allowed increases of 10 per cent for Iceland and 8 per cent for Australia because they were not using all of their carbon allowance.
- In order to achieve their targets, countries could either cut their emissions or trade with other countries in carbon. This means that a country could buy carbon credits from another country. For example, Iceland could sell 2 per cent of its carbon credits to the EU to enable the EU to meet its target of 8 per cent.

December 2007 – Bali conference

Representatives of over 180 countries were present. The result of the meeting was the Bali Roadmap in which initiatives were agreed to try to reach a secure future climate.

Local scale responses to climate change

These are the actions that are being taken by people and local councils to try to play their part in reducing climate change.

By local councils
The UK government believes that local councils can play an important part in the reduction of carbon emissions. The government has given local councils £4m to help them to develop ideas which will cut carbon emissions. Woking Borough Council has used the money to provide power to some of its public buildings, for example Vyne Community Centre has solar panels on the roof which generate electricity.

By schools
'Live simply' is a campaign which ran throughout the whole of 2007. It was initiated by the Catholic Church to encourage students to consider how they make choices in life. It provided a number of resources for schools which made students think about their impact on the world and sustainability.
Many schools are introducing energy efficient water and central heating systems run from renewable sources such as wind turbines or solar panels. Schools also have notices to switch off lights.

Local responses to climate change

By local interest groups
One such group is 'Manchester is my planet'. This group is running a 'pledge campaign' to encourage individuals to reduce their carbon footprint. One of the initiatives is the Green Badge Parking Permit. People who own cars which have been recognised as having low carbon emissions can apply for a Green Badge Parking Permit which allows car owners to buy an annual parking permit for NCP car parks within Greater Manchester at a 25 per cent discount. The permit is valid for twelve months.

Exam practice

Describe the responses to climate change on a local scale. **(4 marks)**

Answers online

Online

Exam tip
If the command word is **describe**, you should state the main characteristics.

Attitudes to climate change

People have different attitudes to climate change. Some deny it is happening while others accept that it is happening but do not believe that it is being caused by humans. Below are a number of statements which reflect this range of attitudes.

Check your understanding — Tested

Read the attitudes to climate change in the spider diagram. Add some more of your own in the blank speech bubbles in the diagram. You could provide counter-arguments to points raised in the diagram.

Go online for answers — Online

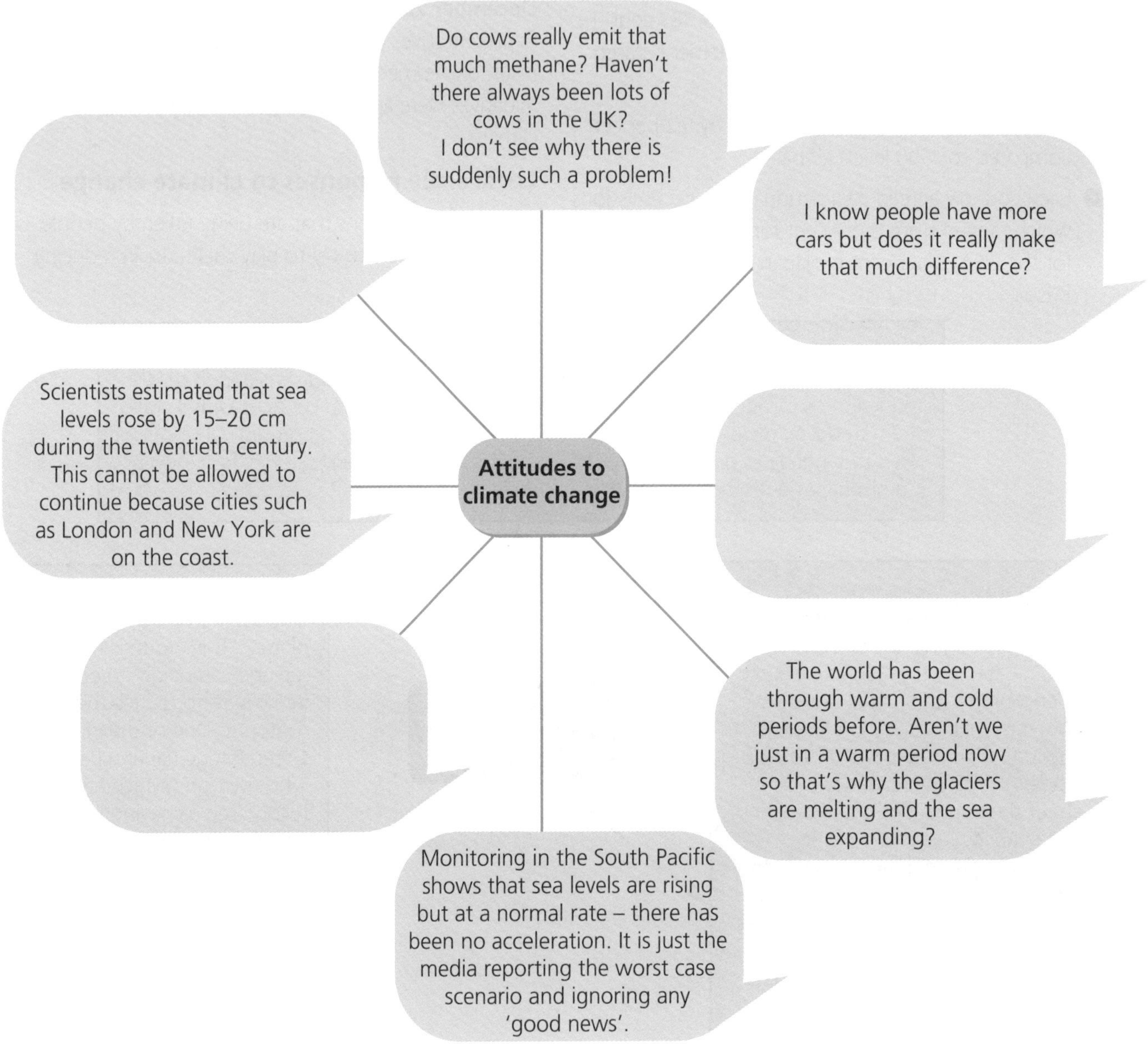

Do cows really emit that much methane? Haven't there always been lots of cows in the UK? I don't see why there is suddenly such a problem!

I know people have more cars but does it really make that much difference?

Scientists estimated that sea levels rose by 15–20 cm during the twentieth century. This cannot be allowed to continue because cities such as London and New York are on the coast.

Attitudes to climate change

The world has been through warm and cold periods before. Aren't we just in a warm period now so that's why the glaciers are melting and the sea expanding?

Monitoring in the South Pacific shows that sea levels are rising but at a normal rate – there has been no acceleration. It is just the media reporting the worst case scenario and ignoring any 'good news'.

Sustainable development for the planet

Definitions and interpretations of sustainable development

Revised

In 1980, the United Nations released the Brundtland report which defined sustainable development as:

> 'development which meets the needs of the present without compromising the ability of future generations to meet their own needs.'

A key area of sustainable development is that it should not hinder development but give a better quality of life both now and in the future. In the UK, four key sustainable areas have been identified.

- **Climate change and energy** – greenhouse gas emissions should be reduced in the UK and worldwide while at the same time countries should prepare for the climate change that cannot be avoided.

- **Natural resources** – the limits of the natural resources that sustain life, such as water, air and soil, are understood so that they can be used most efficiently.

- **Sustainable communities** – places that people live and work in need to be looked after by implementing ideas such as ecotowns and green energy.

- **Sustainable consumption and production** – the ways that products are designed, produced, used and disposed of should be carefully controlled.

The development of policies by large organisations to make them more sustainable

Revised

Large organisations have realised that they must be more sustainable. They can achieve this in many different ways:

- in the manufacturing of the product
- in the recycling of packaging material
- by encouraging customers to recycle products
- by encouraging employees to be more sustainable in the workplace.

Exam practice

Define the term sustainable development. **(2 marks)**

Answers online

Online

The food industry – Asda/Walmart

Asda's distribution centre in Didcot, Oxfordshire, now recycles all of its plastic packaging.

Asda benefits because it is paid for the plastic which is recycled. The environment benefits because plastic is not sent to landfill sites.

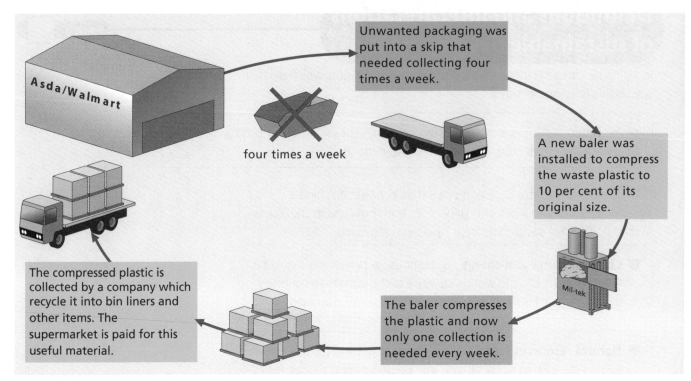

Asda/Walmart

four times a week

Unwanted packaging was put into a skip that needed collecting four times a week.

A new baler was installed to compress the waste plastic to 10 per cent of its original size.

Mil-tek

The baler compresses the plastic and now only one collection is needed every week.

The compressed plastic is collected by a company which recycle it into bin liners and other items. The supermarket is paid for this useful material.

↑ Figure 5 A supermarket's approach to reducing landfill by recycling using a baler to compress waste plastic

The communications industry – Nokia

Nokia are concerned that people are not recycling their old phones. If every mobile phone user recycled one phone it would save 240,000 tonnes of raw materials. Nokia gives information on its website on where to find recycling points and the address to send the phone to if there is not a centre nearby.

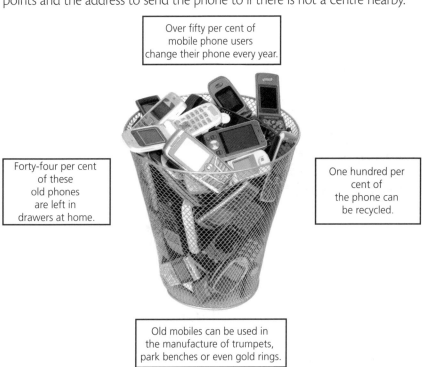

Over fifty per cent of mobile phone users change their phone every year.

Forty-four per cent of these old phones are left in drawers at home.

One hundred per cent of the phone can be recycled.

Old mobiles can be used in the manufacture of trumpets, park benches or even gold rings.

↑ Figure 6 Mobiles about to be recycled

A global company – General Electric

This is a large transnational corporation (TNC) which operates in many different countries. It is trying to produce its products in a more sustainable way. By 2012 it hoped to reduce its fresh water usage by 20 per cent. This is expected to save 7.4 million cubic metres of water which is enough to fill 3000 Olympic-sized swimming pools. This will be achieved by monitoring their water usage and improving their water recycling. Much of the water in their boilers and cooling towers will be recycled water.

Power generation – coal-fired power stations

Coal-fired power stations provide 38 per cent of the world's energy and in countries such as China they provide over 75 per cent. This reliance on coal as an energy source and the resultant pollution means that coal-fired power stations need to be as efficient as possible in order to produce the least amount of pollution. Coal-fired power stations emit large amounts of carbon dioxide (CO_2), sulphur dioxide (SO_2) and nitrous oxide (N_2O). The emissions of these gases are a major contributor to both acid rain and climate change. Figure 8 shows ways that the emissions of these gases are being reduced.

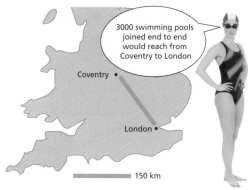

3000 swimming pools joined end to end would reach from Coventry to London

Coventry

London

150 km

↑ **Figure 7 General Electric wants to save 7.4 million cubic metres of water, which is enough to fill 3000 Olympic-sized swimming pools**

Exam tip
- You could be asked questions that require recall of knowledge. You should learn specific points (facts and figures) about the policies of large organisations to make them more sustainable.
- Read the question carefully and don't get caught out.

Carbon dioxide

Amine scrubbers can be fitted into the chimneys of power stations. These use amine solutions to remove CO_2 from the waste gases. In this way, up to 98 per cent of CO_2 can be removed.

Sulphur dioxide

If scrubbers are fitted into the flue of the power station, 95 per cent of SO_2 emissions are removed. In Germany all coal-fired power stations have scrubbers fitted but this is not the case in LICs.

Nitrous oxide

Most power stations have systems fitted that will remove up to 70 per cent of nitrogen emissions.

↑ **Figure 8 A biomass power station in Slough**

Sustainability in the workplace

Many companies have signs up asking their employees to switch off their computers and lights at the end of the day.

Some large companies ask their employees to recycle their rubbish into different bins.

Many workplaces are encouraging their employees to be more sustainable

The internet is being used to send information rather than the post. Many companies aim to become **paperless** in the future.

The use of **video conferencing** reduces a company's **carbon footprint**.

Check your understanding | Tested

How does the internet cut down on paper and cut carbon dioxide emissions?

Go online for answers | Online

Key terms

Video conferencing – using video links so that employees can stay in their offices watching and listening to conferences on their computers rather than travelling many miles to attend conferences.

Carbon footprint – the impact that people have on the planet measured through the amount of carbon that they use for certain activities.

Paperless companies – companies which use email to communicate with people rather than paper and the postal service.

The management of transport in urban areas | Revised

Sustainable transport involves maintaining the standard of transport that is required for society and the economy to function efficiently without placing too much pressure on the environment.

In urban areas of the world there is a great dependency on the car as a means of transport. In both HICs and LICs people are becoming ever more dependent on private vehicles for moving around the city. Car ownership is growing most rapidly in LICs and MICs. In Delhi, India, the number of vehicles in the city has grown from ½ million in 1970 to over 5 million in 2008.

Governments want people to give up using their cars and use public transport more frequently. The problem is that car drivers will not use public transport until it is cheaper and more efficient.

Check your understanding | Tested

What is meant by the public versus the private debate?

Go online for answers | Online

Sustainable transport schemes

There is a range of sustainable transport schemes which can be used including:

● car sharing where workers share lifts to work using their own cars – if half of UK motorists received a lift one day a week, vehicle congestion and pollution would be reduced by 10 per cent

● designated cycle and walking paths within the urban area – Milton Keynes has 273 km of cycle paths

● road lanes that only allow cars with at least two passengers in them

● road lanes which give priority to buses, ensuring they get an easy passage through congested areas – there is a designated bus lane from the A329M right into the centre of Reading.

Congestion charging – the practice of making motorists pay to travel into the centre of large urban areas during periods of heaviest use.

Park and ride – schemes that allow shoppers to park their cars in large designated parking areas on the edge of the urban area and catch a bus into the town centre. Parking is free but there is often a charge for bus travel to the city centre.

London congestion charge

● It costs £8 to enter the central area of London between 7.00a.m. and 6.00p.m., Monday to Friday.

● By 2008 there had been the following improvements:

  ● traffic levels reduced by 21 per cent

  ● 65,000 fewer car journeys a day

  ● 12 per cent increase in cycle journeys within the zone

  ● 12 per cent reduction in the emission of nitrous oxide and fine particulates.

Describe four ways that transport can be managed sustainably.

Cambridge park and ride

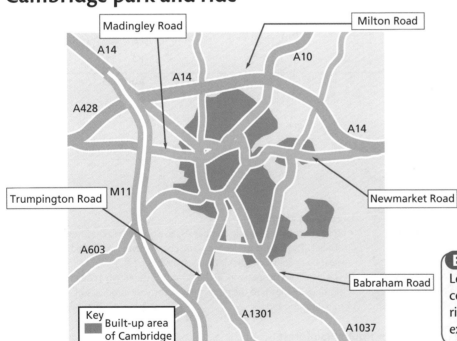

● There are 4500 parking spaces available.

● Double-decker buses carrying up to 70 passengers leave the parks every ten minutes during the day from Monday to Saturday.

● It costs £2.20 for a return ticket to the city centre.

↑ Figure 9 Cambridge park and ride scheme

Learn the definitions of the terms congestion charge and park and ride as well as the details of the examples given.

Make a list of all the specific points (facts and figures) about the Cambridge park and ride scheme and the congestion charge in London.

Using examples, explain how traffic is being managed sustainably in urban areas.　**(4 marks)**

Answers online

Online

The effects of resource extraction from tropical rainforests

Revised

Tropical rainforests are being destroyed at the rate of 32,000 hectares per day. The size of the remaining forest is about 5 per cent of the world's land surface. Much of the area which remains has been affected by human activities and no longer contains its original biodiversity.

Oil extraction in Ecuador
- Miscarriages are common and stomach cancer is five times more frequent in the Oriente region because of hydrocarbons in the river water.
- Many plants such as the periwinkle, which is used to cure childhood leukaemia, are becoming extinct.

Mining in Brazil
- The Carajas iron ore plant uses wood to power the plant. This results in annual deforestation of 6100 km².
- The River Tapajos is contaminated with mercury which is used in gold mining. This affects 90 per cent of all fish caught and leads to high levels of cancer in the local people.

Mining in Indonesia
- Every day 285,000 tonnes of mining waste is dumped into the River Aghawaghon. This pollutes fish and means there is a shortage of water for local people.
- Crocodiles in the area of Teluk Etna are currently on the brink of extinction.

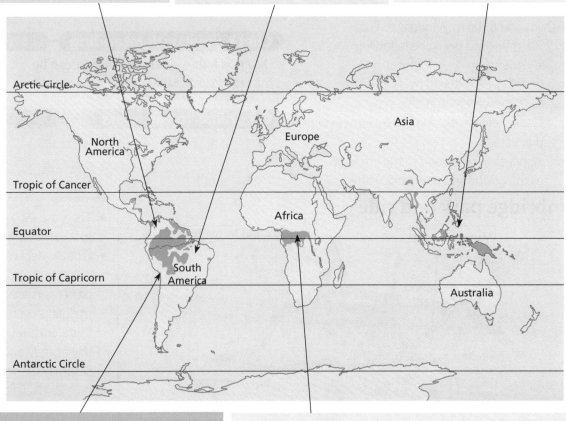

Gas pipeline in Peru
- Local people are exposed to diseases that they are not immune to. During the 1980s half of the Nahua died from influenza and whooping cough.
- Many roads have been built through the forest in the Camisea region allowing settlers into the area who then cut down the forest to farm.

Logging in Cameroon
- Roads built by the logging companies have opened up the forest to hunters. This has led to elephants and chimpanzees being killed and their meat being sold for high prices to restaurants.
- The local Baka people work in the sawmills without any protective clothing. This leads to them breathing in the toxic products which are used to treat the wood.

↑ Figure 10 The effect of resource extraction from tropical rainforest

Exam practice

Explain the effects of resource extraction on rainforest areas.

(6 marks + 4 for SPaG)

Answers online

Online

Exam tip

Read the question carefully. It is easy to become confused between effects and management in rainforest areas.

Tropical rainforest management

Revised

Ecuador

Oil has been extracted from the Amazon rainforest in Ecuador since the 1960s. The companies have done very little to manage the effects of the extraction. Recently the local indigenous people have taken the oil companies to court because of the destruction of their environment.

Texaco have agreed to pay $40 million to cover its share for cleaning up, among other things, some 160 of the 600 waste pits created. But the chief of the local Secoya tribe stated that $6 billion was needed to do the job properly.

Venezuela

Since 2008, the government of Venezuela has not issued any more permits to mine gold or diamonds in the Imataca Rainforest Reserve or anywhere else in the country. Due to its oil reserves, the country does not need to exploit the minerals for economic reasons therefore it can afford to conserve its forest area.

Madagascar

In 2001 Givaudan, a Swiss company, sent a team to Madagascar to survey for new fragrances. It developed 40 aromas that were then sold. The company shared the profits with local communities through conservation and development initiatives.

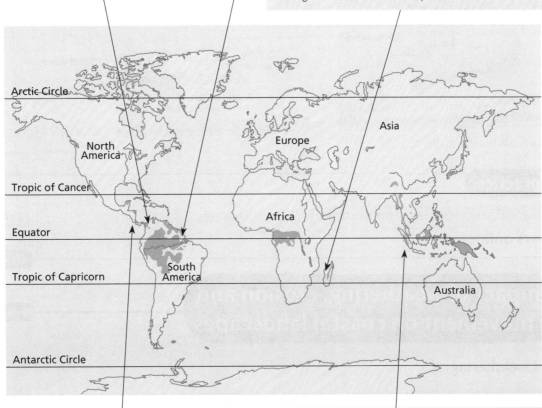

Costa Rica

Costa Rica is developing its rainforest in a sustainable way. One of the ways is through ecotourism. Many areas of the country, including the famous Cloud Forest area, have developed tourist facilities such as zip wiring and trails through the forest which are very popular with the tourists.

Malaysia

In Malaysia the government has rejected plans to build a coal-fired power plant at Silam, on the island of Borneo. The government decided that they did not want to pollute the area and more environmentally friendly forms of energy would need to be found.

The country has vast reserves of coal and other minerals such as gold. The government will not develop these resources at the expense of the rainforest which has many endangered species such as the orang-utan. Instead it is going to develop ecotourism, emphasising the natural attractions such as world-class diving and the biologically diverse tropical rainforest.

↑ Figure 11 Tropical rainforest management

Check your understanding

Tested

Construct a table. The first column should have all the specific points about the effects of resource extraction on rainforest areas. The second column should contain the specific points about rainforest management.

Go online for answers

Online

Coastal processes produce landforms

Types of waves

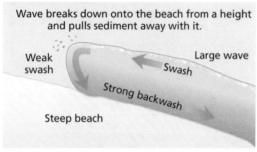

Wave breaks down onto the beach from a height and pulls sediment away with it.

Weak swash

Large wave

Swash

Strong backwash

Steep beach

↑ **Figure 1a A destructive wave**

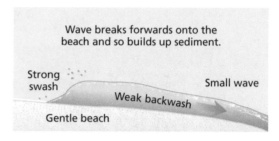

Wave breaks forwards onto the beach and so builds up sediment.

Strong swash

Small wave

Weak backwash

Gentle beach

↑ **Figure 1b A constructive wave**

Exam practice

State two differences between constructive and destructive waves. **(2 marks)**

Answers online

The impact of weathering, erosion and mass movement on coastal landscapes

Types of weathering

Physical weathering: freeze–thaw action

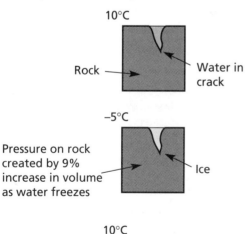

10°C

Rock → Water in crack

−5°C

Pressure on rock created by 9% increase in volume as water freezes → Ice

10°C

Crack increased in size

Biological weathering

Seed falls into crack
↓
Rain causes seedling to grow
↓
Roots force their way into cracks
↓
As the roots grow they break up the rock

Burrowing animals also break up rock.

Chemical weathering

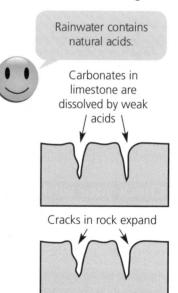

Rainwater contains natural acids.

Carbonates in limestone are dissolved by weak acids

Cracks in rock expand

↑ **Figure 2 Main types of weathering**

Types of erosion

Erosion is when rocks are worn down by an agent of erosion such as water. There are four types of erosion which take place in the sea:

- **Hydraulic action**, **corrasion** and **corrosion** are ways in which the sea erodes cliffs.
- **Attrition** is the breaking up of rocks and pebbles in waves. Look at the key terms box below for detailed definitions of these terms.

Key terms

Weathering – ways that rocks are broken down in situ.

Hydraulic action – the compression of air in cracks puts pressure on the rock and causes pieces of the rock to break off.

Corrasion (abrasion) – sand and pebbles carried in waves are thrown against the cliff face.

Corrosion (solution) – chemicals in sea water dissolve certain rock types, such as chalk.

Attrition – the breaking up of rocks and pebbles in the waves. The movement of waves means that rocks are continually knocked against each other, removing any sharp edges, to produce smooth pebbles and, eventually, sand.

Transportation – the movement of sand and pebbles by the sea.

Deposition – the putting down of sand and pebbles by the sea.

Mass movement – material moves down a slope, pulled down by gravity.

Exam tip

Learn all of the terms in the key term box as you could be asked to define them in an exam.

Exam tip

Remember, attrition is not a type of erosion used to erode cliffs.

Check your understanding

Describe three processes used by the sea to erode cliffs.

Go online for answers

Tested

Online

Types of mass movement

There are many different types of **mass movement**. You need to know about soil creep and slumping.

Soil creep	Slumping
• Slowest downhill movement • Gravity pulls water in the soil downhill • Soil particles move with the water • Heavy rainfall causes faster downhill movement • The slope appears to have ripples • The ripples are known as terracettes	• A large area of land moving down a slope • Common on clay cliffs • Dry weather makes the clay contract and crack • When it rains, water gets into the cracks • The soil becomes saturated • A large piece of rock is pulled down the cliff face • It has slipped on the slip plane of saturated rock

Check your understanding

State three differences between soil creep and slumping.

Go online for answers

Tested

Online

Exam practice

Explain one type of mass movement. **(3 marks)**

Answers online

Online

Landforms created by coastal erosion

Erosional features – cliffs and wave-cut platforms

In these boxes the explanation is underlined.

Above the wave-cut notch an overhang develops. As the notch becomes larger the overhang will become unstable. This is because of its weight and the lack of support. In time the overhang will fall due to the pull of gravity.

As the width of the platform increases the power of the sea decreases, because it has further to travel to reach the cliff and the water is shallower causing more friction.

In this box the process explanation is in bold.

The cliff is eroded at the bottom by **corrasion. This is pebbles carried by the sea which are thrown against the cliff by the breaking wave, knocking off parts of the cliff.** In time a wave-cut notch is formed.

High water mark

Low water mark

The sea continues to attack the cliff in this way and the cliff retreats.

The remains of the cliff, now below the sea at high tide, form a rocky wave-cut platform. The platform will also contain the boulders which have fallen from the cliff.

Exam practice

Explain the formation of cliffs and wave-cut platforms. **(4 marks)**

Answers online

Online

Erosional features – headlands and bays

Headlands and bays form due to different rock types.

● They only occur on coastlines where soft and hard rocks are found at right angles to the sea.

● The soft rock erodes more quickly than the hard rock, forming bays.

● The hard rock is more resistant and sticks out as headlands.

● You should also refer to the processes of erosion.

Exam tip

A grade candidates may also refer to the fact that erosion eventually becomes greater on the headlands because the bays have retreated and the headlands are more exposed.

The formation of caves, arches, stacks and stumps

This is just description.

A stump is formed by the action of the sea and weathering. The sea erodes a crack using hydraulic action and makes it bigger, forming a cave. If it is a headland, caves will form on either side. Eventually the backs of the caves meet and an arch is formed. In time the arch will collapse forming a stack. The stack will then collapse forming a stump.

Below is an A* grade answer that explains how a stump is formed. Note the comments, which pick out the **sequence description, explanation** and the **processes.**

| Sequence description | → | A stump is formed by the action of the sea and weathering. |

| Process explanation | → | The sea erodes a crack with hydraulic action. This is when water hits the cliff compressing air in cracks. As the water retreats the pressure is released breaking off pieces of rock. |

| Sequence description | → | This makes the crack bigger, forming a cave. If it is a headland, caves will form on either side. Eventually the backs of the caves meet and an arch is formed. In time the arch will collapse, forming a stack. |

| Process explanation | → | This is due to undercutting of the sea. One of the processes the sea uses is corrasion which is when rocks in the sea are thrown against the cliff breaking off pieces of the cliff. |

| Sequence explanation | → | The arch becomes wider at the bottom and is unable to support the weight above. Eventually the arch is pulled down by gravity. |

| Sequence and process | → | Weathering is also active on the cliff face and further weakens the cliff. Through time the sea erodes the bottom of the stack and it collapses to leave a stump, which is covered by the sea at high tide. |

This is a C grade answer. Again the **sequence description, explanation** and **processes** are picked out.

Process
Sequence described
Sequence explained

A stack is formed by the action of the sea. The sea erodes a crack in the cliff using corrasion. Eventually the cave becomes an arch. In time the arch will collapse forming a stack. This is due to undercutting of the sea. The arch becomes wider at the bottom and is unable to support the weight above. Eventually the arch is pulled down by gravity.

Check your understanding
Tested

Explain the formation of headlands and bays. After you have completed your answer, underline the process explanation in one colour and the sequence explanation in another.

Go online for answers
Online

Exam tip

- **Foundation Tier** – For questions about landform formation, describe what happens like a sequence of events. Don't forget to explain part of the sequence and don't forget the process.
- **Higher Tier** – For questions about landform formation, describe in detail and explain what happens in a sequence of events. Don't forget to fully explain the processes involved.

Exam tip

- Always include a process explanation when you are answering questions about landform formation.
- A diagram will always help your answer.

Landforms created by coastal deposition

What is longshore drift?

Longshore drift is the movement of sediment along the shore. Pebbles are moved in the direction of the prevailing wind. This can lead to the formation of spits and bars. It can be a problem in river estuaries due to the deposition of sediment. Some harbours, such as Poole Harbour, have to be constantly dredged due to longshore drift occurring at their entrance.

Exam tip

You could be asked to describe the process of longshore drift or annotate a half-completed diagram. Don't forget to **explain** as well as **describe** what is occurring.

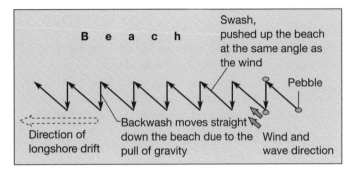

Swash, pushed up the beach at the same angle as the wind

B e a c h

Pebble

Direction of longshore drift

Backwash moves straight down the beach due to the pull of gravity

Wind and wave direction

Check your understanding

What is the impact of longshore drift on the coastline?

Go online for answers

Depositional features – beaches

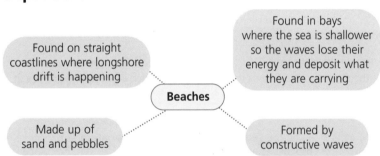

Found on straight coastlines where longshore drift is happening

Found in bays where the sea is shallower so the waves lose their energy and deposit what they are carrying

Beaches

Made up of sand and pebbles

Formed by constructive waves

Depositional features – spits and bars

● Spits are narrow stretches of sand and pebbles that are joined to the land at one end.

● If you are asked to explain the formation of a bar you need to discuss spit formation and then add the following points:

  ● Bars are spits which go across a bay.

  ● This is only possible if there is shallow water and no river entering the sea.

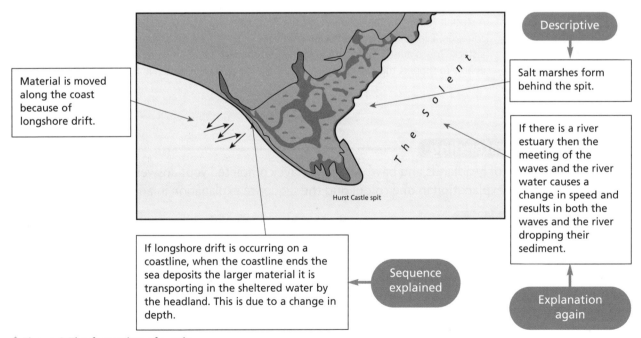

Material is moved along the coast because of longshore drift.

Hurst Castle spit

The Solent

Descriptive

Salt marshes form behind the spit.

If there is a river estuary then the meeting of the waves and the river water causes a change in speed and results in both the waves and the river dropping their sediment.

If longshore drift is occurring on a coastline, when the coastline ends the sea deposits the larger material it is transporting in the sheltered water by the headland. This is due to a change in depth.

Sequence explained

Explanation again

↑ **Figure 3 The formation of a spit.**

Hint

You have already learnt the processes. Look back to pages 38–39 in this revision guide.

Coastal landforms are subject to change

What affects the rate of coastal erosion?

Revised

The rate at which a cliff recedes depends on processes including erosion, weathering and mass movement. The rate at which these processes happen depends on **fetch**, **geology** and coastal management.

Fetch

- The longer the fetch the stronger the wave.

Geology

- Rock type – harder rocks are eroded more slowly, for example granite. Think of headlands and bays!
- Rock angle to the coast – parallel or at right angles.

Coastal management

If a coastline is defended then it will erode much more slowly.

- Think of Walton (see table on page 47).

> **Key terms**
>
> **Fetch** – the distance that wind travels over open water.
>
> **Geology** – the rock type and structure of an area.
>
> **Coastal recession** – where the coastline is eroded so that it moves back.

What are the effects of coastal recession on people and the environment?

Revised

The effects of **coastal recession** should be studied through a range of examples.

A Doomed Village! Happisburgh

Since 1995, 25 properties and the village's lifeboat launching station have been lost to the sea. The village contains eighteen listed buildings including a Grade 1 listed church which is estimated to be in the sea by 2020. The life of the villagers is totally dominated by their struggle against the sea.

Tower's days are numbered!

The Tower at Walton-on-the-Naze could soon be lost to the sea if the cliff continues to erode at its present rate of 1.5 m a year. The area around the Tower is used for recreational purpose and is not deemed worthy of coastal protection.

Train passengers get a shower!

Passengers on the train travelling from Exeter to Plymouth and Penzance regularly get a shower as the sea washes over the tracks. On one occasion, 160 passengers were stranded on a train for four hours while the sea washed over them because the train's electrics were not working.

Golfers' paradise threatened!

A number of golf courses around the country are losing precious greens and fairways to the sea.

- Sheringham golf course in Norfolk is soon to lose its fifth and sixth holes.
- The Royal North Devon Golf Club at Westward Ho! is losing its seventh and eighth holes.

Barton on Sea becomes Barton in the sea!

Since 1975 Barton has lost the following properties to the sea:

- seaside café demolished because it had become dangerous
- Manor Lodge demolished before it fell into the sea
- 2004 coastal footpath closed; re-sited further back from cliff edge.

> **Exam practice**
>
> Explain the effects of coastal recession. Use examples in your answer. **(4 marks)**
>
> **Answers online**
>
> Online

How are the effects of coastal flooding reduced by prediction and prevention?

This should be studied through **forecasting**, **building design**, **planning** and **education**.

Forecasting
- The Met Office predicts (forecasts) the likelihood of a flood. The information gets to householders through weather forecasts and news broadcasts on the TV and radio. It is also on their website.
- The Environment Agency monitors sea conditions 24 hours a day, 365 days a year. This Storm Tide Forecasting Service provides forecasts of **coastal flooding**. Information is provided on a 24-hour flood hotline and the Environment Agency website.

Building design
- In Bangladesh all one-storey and two-storey buildings must have an external staircase to the roof.
- Houses along the coast at Malibu, near Los Angeles in California, are built on stilts to protect them from storm tides.
- Houses should minimise penetration from wind, rain and storms.

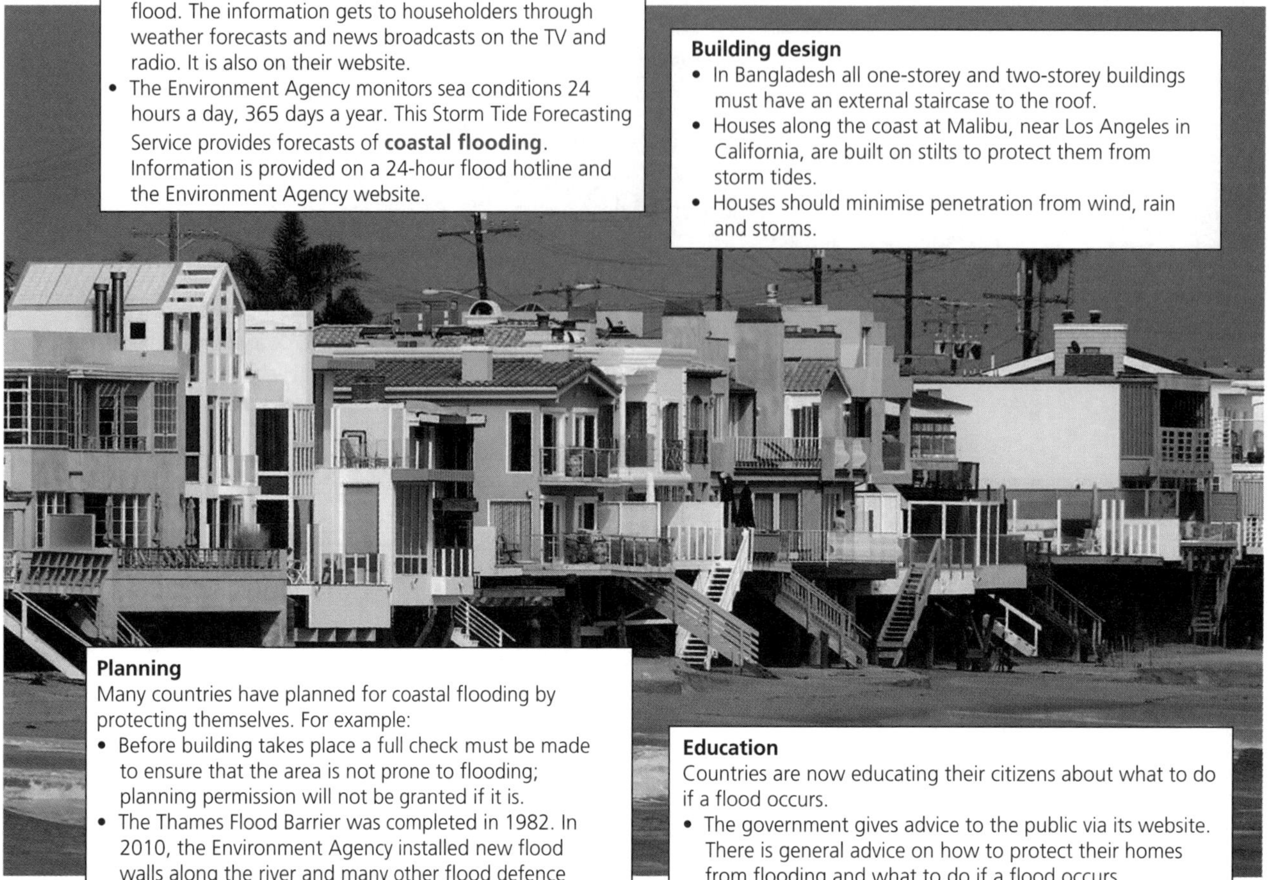

Planning
Many countries have planned for coastal flooding by protecting themselves. For example:
- Before building takes place a full check must be made to ensure that the area is not prone to flooding; planning permission will not be granted if it is.
- The Thames Flood Barrier was completed in 1982. In 2010, the Environment Agency installed new flood walls along the river and many other flood defence techniques to protect the London area against coastal flooding.
- In Bangladesh the Coastal Embankment project has led to the building of twelve sea-facing flood walls and 500 flood shelters.

Education
Countries are now educating their citizens about what to do if a flood occurs.
- The government gives advice to the public via its website. There is general advice on how to protect their homes from flooding and what to do if a flood occurs.
- In Bangladesh many coastal areas have flood warning systems.
- In King's Lynn in Norfolk there is a flood siren and people are employed by the council to go from house to house to warn people and help them to prepare.

↑ **Figure 4 Houses along the coast of Malibu**

Check your understanding

Are these statements true? Give reasons for your answer.

- Floods can be predicted with adequate forecasting.
- Floods can be prevented by adequate building design, planning and education.

Go online for answers

Key terms

Prediction – a forecast of what might happen.

Prevention – trying to stop something happening.

Building design – the style of a building which helps to prevent flooding.

Coastal flooding – the inundation by the sea of areas near to the coast.

What are the main types of hard and soft engineering used on the coastline of the UK?

Revised

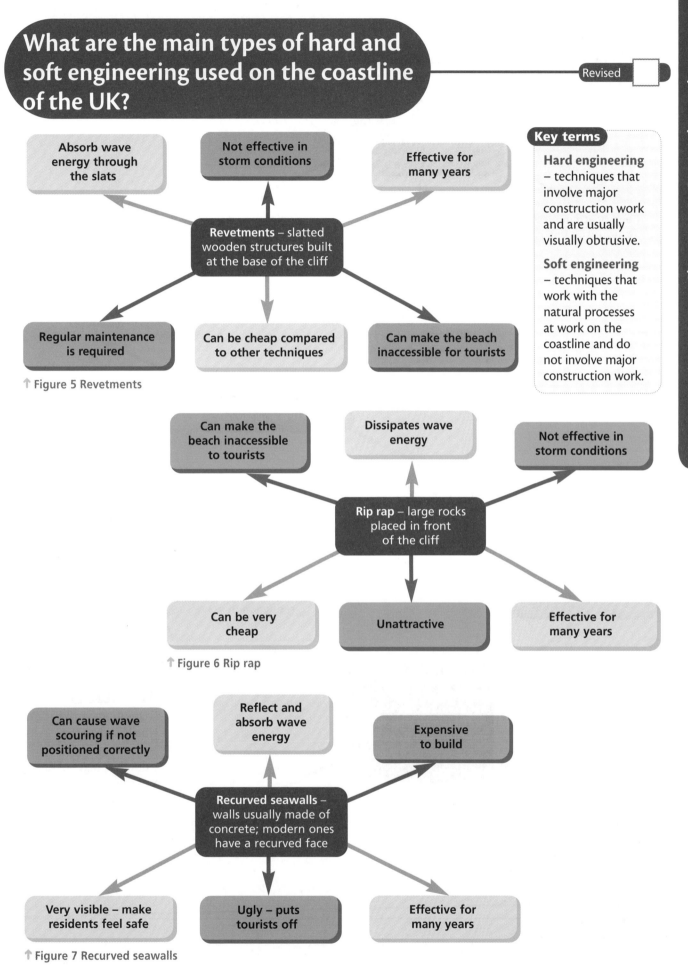

Absorb wave energy through the slats

Not effective in storm conditions

Effective for many years

Revetments – slatted wooden structures built at the base of the cliff

Regular maintenance is required

Can be cheap compared to other techniques

Can make the beach inaccessible for tourists

↑ Figure 5 Revetments

Key terms

Hard engineering – techniques that involve major construction work and are usually visually obtrusive.

Soft engineering – techniques that work with the natural processes at work on the coastline and do not involve major construction work.

Can make the beach inaccessible to tourists

Dissipates wave energy

Not effective in storm conditions

Rip rap – large rocks placed in front of the cliff

Can be very cheap

Unattractive

Effective for many years

↑ Figure 6 Rip rap

Can cause wave scouring if not positioned correctly

Reflect and absorb wave energy

Expensive to build

Recurved seawalls – walls usually made of concrete; modern ones have a recurved face

Very visible – make residents feel safe

Ugly – puts tourists off

Effective for many years

↑ Figure 7 Recurved seawalls

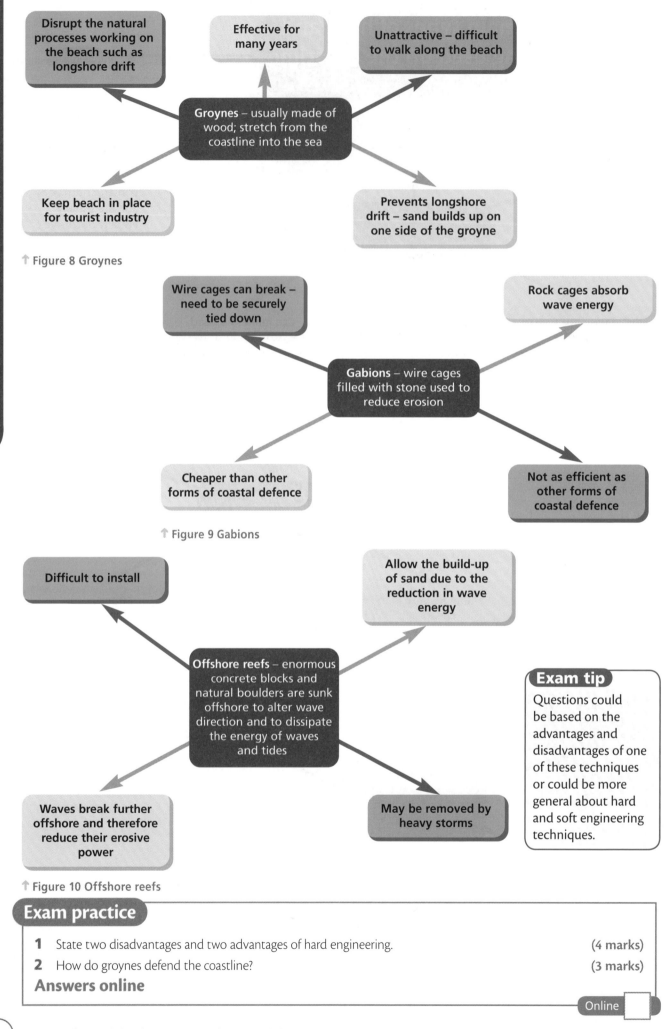

Disrupt the natural processes working on the beach such as longshore drift

Effective for many years

Unattractive – difficult to walk along the beach

Groynes – usually made of wood; stretch from the coastline into the sea

Keep beach in place for tourist industry

Prevents longshore drift – sand builds up on one side of the groyne

↑ Figure 8 Groynes

Wire cages can break – need to be securely tied down

Rock cages absorb wave energy

Gabions – wire cages filled with stone used to reduce erosion

Cheaper than other forms of coastal defence

Not as efficient as other forms of coastal defence

↑ Figure 9 Gabions

Difficult to install

Allow the build-up of sand due to the reduction in wave energy

Offshore reefs – enormous concrete blocks and natural boulders are sunk offshore to alter wave direction and to dissipate the energy of waves and tides

Waves break further offshore and therefore reduce their erosive power

May be removed by heavy storms

Exam tip

Questions could be based on the advantages and disadvantages of one of these techniques or could be more general about hard and soft engineering techniques.

↑ Figure 10 Offshore reefs

Exam practice

1 State two disadvantages and two advantages of hard engineering. (4 marks)
2 How do groynes defend the coastline? (3 marks)

Answers online

Online

Soft engineering technique	Description	Advantages	Disadvantages
Beach replenishment	The placing of sand and pebbles on a beach	• Looks natural • Provides a beach for tourists • A beach is the best form of natural defence because it dissipates wave energy • Cheap	• May affect plant and animal life in the area • Requires constant maintenance; can all be washed away very quickly in as little as one year • Disruption for homeowners – large noisy lorries full of sand regularly replenish the beach
Cliff regrading	The cliff is cut back and given a new gentle slope to stop it slumping	• May be covered in ecomatting to encourage vegetation growth • Very natural – will encourage wildlife in the area	• Not effective alone – needs other defences at the cliff foot • Some homes on the cliff may have to be demolished
Managed retreat	Allowing the sea to gradually flood land or erode cliffs	• Creates new habitats for plants and birds • Cheap	• Upsetting for landowners who lose land • Difficult to estimate the extent of sea movement, especially with rising sea levels

↑ Figure 11 Soft engineering techniques

Check your understanding
Tested

What are the differences between hard and soft engineering techniques?

Go online for answers
Online

Exam tip

Read the question carefully and don't get caught out. Underline the key words in the question.

Coastal management

How the coast is managed in a named location
Revised

You need to be able to describe and explain how the coast is managed at Walton-on-the-Naze.

Year	Type of management
1977	Groynes built to stop longshore drift movement from south to north Cliff regraded and drainage channels installed to produce a gentle more stable slope Slope planted with gorse and nettles to stop people climbing on the cliff Seawall built to protect the soft London clay at the bottom of the cliff
1998	300 tonnes of Leicester granite placed around the Tower breakwater
1999	Beach replenished with sand and gravel from Harwich Harbour

Exam tips

• You have already learnt the effects of coastal recession on Walton. Be careful not to confuse management with effects.
• Learn the specific points such as the year for each type of management.
• Look back at the mark schemes for questions on case studies on page viii of the introduction.

Check your understanding
Tested

Use the table to list the specific information (facts and figures) about the coastal defences which have been built at Walton-on-the-Naze.

Go online for answers
Online

Exam practice

For a named example, explain how the coast is being managed. **(4 marks)**

Answers online
Online

Chapter 8 River Landscapes

River processes produce distinctive landforms

Drainage basin terms
Revised

Key terms

Mouth – where a river meets the sea.

Source – the start of a river.

Tributary – a stream that joins a larger river.

Watershed – the boundary of a river basin.

Confluence – the point where two rivers meet.

Exam tip

Learn the definitions of the terms in the key terms box and make sure you can label them on diagrams of river basins.

The impact of weathering, erosion and mass movement on river landscapes
Revised

Types of weathering

Physical, biological and chemical weathering also have an impact on river landscape. Look back at page 38 to revise these processes.

Types of erosion

Erosion is when rocks are worn down by an agent of erosion such as water. There are four types of erosion which take place in rivers.

- **Hydraulic action**, **corrasion** and **corrosion** are ways in which the river erodes its bed and banks.

- **Attrition** is the breaking up of rocks and pebbles in the river water. Look at the key terms box on the right for detailed definitions of these terms.

Key terms

Weathering – ways that rocks are broken down in situ.

Hydraulic action – the pressure of water against the banks and bed of the river. It also includes the compression of air in cracks: as the water gets into cracks in the rock, it compresses the air in the cracks; this puts even more pressure on the cracks and pieces of rock may break off.

Corrasion – particles being carried by the river are thrown against the river banks.

Corrosion (solution) – a chemical reaction between certain rock types and the river water.

Attrition – the breaking up of sediment being carried by the river. Stones and pebbles are continually knocked against each other by the water causing them to become smoother and smaller.

Deposition – when a river drops some of the sediment that it is carrying.

Mass movement – material moves down a slope pulled down by gravity.

Exam tip

- Learn all of the terms in the key term box as you could be asked to define them in an exam.
- Remember to use these terms when explaining how river landforms are formed.

Check your understanding
Tested

Describe three processes used by rivers to erode their bed and banks.

Go online for answers
Online

Exam practice

1 Define the term watershed. **(2 marks)**
2 Explain the process of biological weathering. **(3 marks)**

Answers online

Online

Types of mass movement

There are many different types of **mass movement**. You need to know about soil creep and slumping.

Soil creep	Slumping
• Slowest downhill movement • Gravity pulls water in the soil downhill • Soil particles move with the water • Heavy rainfall causes faster downhill movement • The slope appears to have ripples • The ripples are known as terracettes.	• This happens on river banks; part of the bank slips into the river • Common when the river passes through areas of clay • Dry weather makes the clay contract and crack • When it rains, water gets into the cracks and is absorbed making the rock heavy • The soil becomes saturated • The rock is weakened and slips into the river due to the pull of gravity • The rock has slipped on the slip plane of saturated rock.

Check your understanding   Tested

How does slumping occur on river banks?

Go online for answers   Online

Change in the characteristics of a river   Revised

As the river moves from its source to meet the sea at its mouth, its characteristics will change.

Width	The river will become wider as the amount of water in it grows due to it being joined by tributaries as it moves towards the sea.
Depth	The river will become deeper as the amount of water in it grows due to it being joined by tributaries as it moves towards the sea.
Velocity	The speed of the river is slower in the hills than near to the sea. This is because, although there is a steep gradient in the hills, much of the water is in contact with the bed and banks causing friction. This slows the river down. As it moves towards the sea and becomes deeper and wider, less water is in contact with the bed and banks therefore less friction occurs and the velocity increases.
Discharge	The discharge increases as the river moves towards the sea because of the increase in volume due to it being joined by tributaries.
Gradient	The slope will become less steep as the river moves out of the hills and into flatter areas on its way to the sea.

Key terms

Width – how far it is from one bank of a river to the other.

Depth – the distance between the top and the bottom of the water in a river.

Velocity – the speed that the water in the river is moving at.

Discharge – the volume (amount of water in the river) times the velocity of the river.

Gradient – the slope that the river is moving down.

Exam tip

You could be asked to annotate a diagram which explains the change in the characteristics of a river as it moves from its source to its mouth.

Exam practice

Explain one type of mass movement.   **(3 marks)**

Answers online

Online

Landforms found in river valleys

Revised

The processes at work in a river channel form a number of different landforms.

Waterfall formation

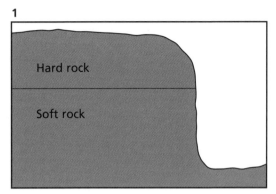

1

Hard rock

Soft rock

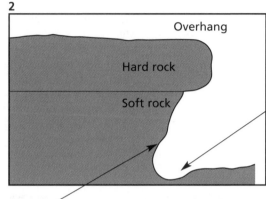

2

Overhang

Hard rock

Soft rock

Plunge pool formed by the force of the water. It is deepened by the process of/ corrasion as sediment being carried by the water is scraped against the bottom and sides.

Soft rock eroded back more quickly (by splash back) caused by hydraulic action

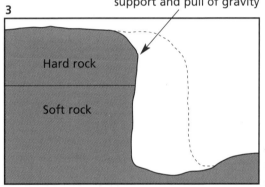

Overhang has fallen due to lack of support and pull of gravity

3

Hard rock

Soft rock

Waterfall retreats upstream to leave a gorge

> **Exam tip**
> Always include a process explanation when you are answering questions about landform formation.

Interlocking spurs

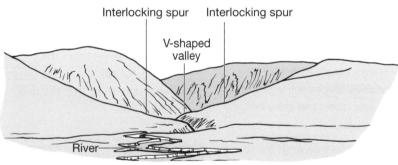

Interlocking spur   Interlocking spur

V-shaped valley

River

Interlocking spurs are barriers of hard resistant rock, which the river cannot easily erode. The river weaves its way around them.

> **Exam tip**
> - **Foundation Tier** – For questions about landform formation, describe what happens like a sequence of events. Don't forget to explain part of the sequence and don't forget the process.
> - **Higher Tier** – For questions about landform formation, describe in detail and explain what happens in a sequence of events. Don't forget to fully explain the processes involved.

> **Exam tip**
> - A grade students should also explain how meander migration aids the formation of a floodplain.
> - A diagram will always help your answer.

Meander bends and ox-bow lakes

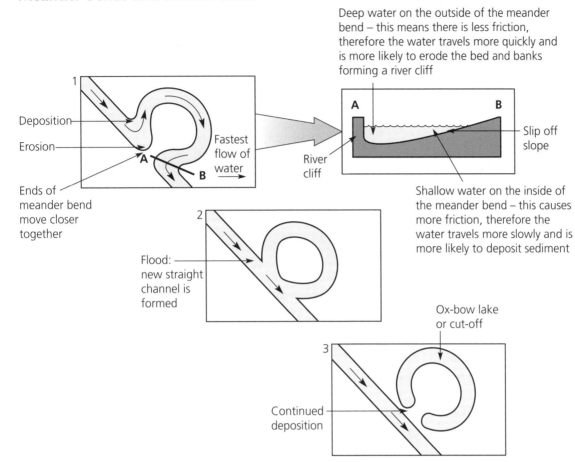

Deep water on the outside of the meander bend – this means there is less friction, therefore the water travels more quickly and is more likely to erode the bed and banks forming a river cliff

Deposition

Erosion

Fastest flow of water

Ends of meander bend move closer together

Flood: new straight channel is formed

River cliff

Slip off slope

Shallow water on the inside of the meander bend – this causes more friction, therefore the water travels more slowly and is more likely to deposit sediment

Ox-bow lake or cut-off

Continued deposition

Floodplain

- This is an area of flat land on either side of the river.
- When there is too much water in the river, this area will flood as the water moves out of the river channel onto the land that surrounds it.
- The water is shallower on the land than it is when it is in the river channel. Therefore, there is more friction and the water drops the sediment it is carrying.
- The water drops the heaviest material first on the banks; the lighter material such as silt is carried the furthest. The deposit of this material forms a floodplain.
- The migration of meanders across a river valley also adds to the floodplain.

Levee

- This is a high bank at the side of a river which is built up during times of flood.
- Each time a river floods, it deposits sediment on its banks. This is because of the change in speed of the water between when it is in the channel and when it has moved out onto the floodplain.
- Over time these banks build up to form levees, which make it harder for the river to flood.

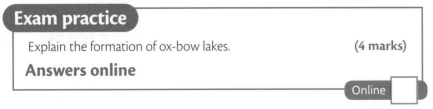

Exam practice

Explain the formation of ox-bow lakes.   **(4 marks)**

Answers online

Online

Flooding and flood prevention

The physical and human causes of river flooding — Revised

When there is more water in an area than a river channel can hold, it will burst its banks. This is known as flooding and is caused by a number of factors.

If a valley has steep slopes, water will move into the river more quickly causing frequent floods.

If there is rain for a number of days, the ground becomes **saturated**. Excess water will then flow straight into rivers causing them to flood.

Physical causes

If an area is formed from **impermeable rocks**, it is more likely to flood more often.

If there has been a heavy snowfall then a sudden rise in temperature, a rapid thaw can happen. Rivers in the area will be unable to cope with the amount of water and will flood.

Key terms

Physical factors – causes of river flooding that relate to the natural environment.

Human factors – causes of river flooding that relate to people.

Saturated – when the ground becomes so wet that it can no longer hold any more water.

Impermeable rocks – rocks that do not allow water through them.

Vegetation – the plants of an area.

Interception – vegetation stopping water from reaching rivers.

Deforestation – when trees and other plants are removed from an area.

Urbanisation – when towns and cities are built.

Exam tip

In an exam question, you may be asked for just **physical factors** or just **human factors** so be sure that you know which are which.

Check your understanding — Tested

Explain how removal of **vegetation** could be classed as a human and a physical cause of river flooding.

Go online for answers — Online

If vegetation is removed from slopes, less **interception** occurs and water will move to the river more quickly.

In extreme cases a dam could burst which would mean there was more water in the river channel and large areas of land close to the river would be flooded.

Human causes

If farmers plough up and down slopes instead of around the hillside, the channels created by the plough allow rainwater to travel faster to the river.

When an urban area is built, there are large areas of tarmac. Rainwater cannot soak into the ground. It runs into drains which allow the water to move into the river at a greater speed. It is therefore more likely to flood.

Check your understanding — Tested

Explain how **urbanisation** and **deforestation** cause rivers to flood.

Go online for answers — Online

The effects of river flooding on people and the environment

Revised

When a river floods, it affects **people** because it can destroy their homes. It also **affects the environment** as land is covered in water, killing animals and crops. Here are some examples.

Bulgaria 2006 – The River Danube flood

- 2 million people were affected by the flood and twenty were killed.
- Roads were impassable; bridges and railway lines were washed away.
- The damage to the economy was estimated at £346m.
- Large areas of farmland were destroyed.

Kenya 2007 – The River Tana flood

- The town of Garissa was under water.
- Roads were impassable and bridges were destroyed.

Mexico 2007 – The River Grijaiva flood in the state of Tabasco

- Approximately one million people were affected by the flood.
- 70 per cent of the state was under water and all crops were destroyed.
- Sandbags were placed around an archaeological site at La Venta.

India 2008 – The River Kosi flood

- 2.7 million people were affected by the flood and 55 were killed.
- Roads were impassable and bridges were destroyed.
- All crops were destroyed.
- The road linking Saharsa village to the rest of the area was washed away.

Exam practice

Explain the effects of river flooding. Use examples in your answer.

(4 marks)

Answers online

Online

Exam tip

Remember to learn specific points (facts and figures) for the effects on people and the effects on the environment.

Check your understanding

Tested

Create a table of the specific points from these examples of flooding that you need to remember. Write points for the effects on **people** in one colour and points for the effects on the **environment** in another.

Go online for answers

Online

Prediction and prevention of the effects of river flooding

This should be studied through forecasting, **building design**, planning and education.

Education
- Governments give advice on how to protect homes from floods via the internet.
- The Environment Agency website gives information on the likelihood of a flood. There is a system of warning codes: flood watch, flood warning, severe flood warning and all clear. These warning codes give people information on what to expect and how to react.

Forecasting
- The Met Office predicts (forecasts) the likelihood of a flood on the TV, radio and internet.
- Householders are advised to either ring a flood hotline number or go onto the Environment Agency website to check for the likelihood of a flood in their area.

Planning
- A flood risk assessment has to be carried out in certain areas before planning permission can be granted.
- A new law was passed in 2010 which requires all new housing in flood risk areas to be resistant to flooding.
- Defra (The Department for Environment, Food and Rural Affairs) has the responsibility for deciding which areas are going to be defended against the risk of river flooding. It also provides most of the funding. The Environment Agency then organises the building and maintenance of the defences.

Building design
Houses can be protected from flooding by using the following techniques.
- Electricity sockets are placed half way up the walls.
- Doors are lightweight and can be moved upstairs.
- Concrete floors are laid instead of wooden ones so they do not rot if they are wet.
- Yacht varnish is used on wooden skirting boards to protect them from water.
- Property is built on stilts such as the one in the photograph.

Check your understanding

Are these statements true? Give reasons for your answer.
- Floods can be predicted with adequate forecasting.
- Floods can be prevented by adequate building design, planning and education.

Go online for answers

Key terms
Prediction – a forecast of what might happen.

Prevention – trying to stop something happening.

River flooding – the inundation by rivers of areas close to their channel.

Building design – the style of a building which helps to prevent flooding.

The main types of hard and soft engineering used to control rivers in the UK

Revised

Hard engineering technique	Description	Advantages	Disadvantages
Flood relief channels	The channel course of the river can also be altered, diverting floodwaters away from settlements.	• No disruption is caused to residents next to the original course of the river • Makes the people who live close to the main river safer as the flood water is diverted into the relief channel • Can be used for water sports • Very effective, should last for many years.	• Extremely expensive • Requires a large amount of land which might be difficult to purchase particularly if it is productive farmland.
Embankments	These are raised banks along the sides of a river.	• Can be used as pedestrian paths beside the river • Earth embankments provide habitats for plants and animals • Concrete embankments are effective at stopping bank erosion.	• Often not built high enough • Concrete embankments are ugly and spoil the view.
Channelisation	A river channel may be widened or deepened allowing it to carry more water; may be straightened so that water can travel faster along the course.	Effectively protects immediate area because water is moved away quickly – long lasting.	• Altering river channel may lead to a greater risk of flooding downstream as the water is carried there faster • Unnatural and visually intrusive.
Dams	Dams are often built along the course of the river in order to control the amount of discharge; water is held back by the dam and released in a controlled way.	The water that is stored in a reservoir behind the dam can be used to generate hydroelectric power or for recreation purposes.	• Sediment is often trapped behind the wall of the dam, leading to erosion further downstream • Settlements and agricultural land may be lost when the river valley is flooded to form a reservoir.

↑ Figure 1 Hard engineering techniques

Key terms

Hard engineering techniques – involve major construction work and are usually visually obtrusive.

Soft engineering techniques – work with natural processes and do not involve major construction work.

Exam tip

Questions could be based on the advantages and disadvantages of one of these techniques or could be more general about **hard** and **soft engineering techniques**.

Soft engineering technique	Description	Advantages	Disadvantages
Floodplain zoning	Local authorities and the national government introduce policies to control urban development close to or on the floodplain.	• Sustainable because it reduces the impact of flooding and building damage is limited • No building on floodplain means less surface run-off.	• Resistance to restricting developments in areas where there is a housing shortage • Enforcing planning regulations and controls may be harder in LICs.
Washlands	The river is allowed to flood naturally in wasteland areas to prevent flooding in other areas.	• Provides potential wetlands for birds and plants • The deposited silt may enrich the soil, turning the area into agricultural land.	• Large areas of land are taken over and cannot be built on • Productive land can be turned into marshland.
Warning systems	A network of sirens give people early warning of possible flooding.	• Cheap • Electronic communication is a very effective way of informing people.	• Sirens can be vandalised • There might not be enough time for residents to prepare.
Afforestation	Trees are planted in the catchment area of the river to intercept the rainfall and slow down the flow of water to the river.	• Relatively cheap • Soil erosion is avoided as trees prevent rapid run-off after heavy rainfall.	• Can make the soil acidic depending upon tree chosen • Increases fire risks because of leisure activities in the forest.

↑ Figure 2 Soft engineering techniques

Check your understanding — Tested

What are the differences between hard and soft engineering techniques?

Go online for answers — Online

Exam practice

1 State two disadvantages and two advantages of soft engineering. **(4 marks)**

2 How do flood relief channels defend urban areas? **(3 marks)**

Answers online — Online

River management

How a river is managed in a named location — Revised

You need to be able to describe and explain how the River Nene has been managed in the Northampton area.

Year	Type of management
2002	The level of the land at Weedon was raised by 10 m by building an embankment across the river valley. Water would be stored behind the embankment in times of flood.
2003	The flood warning system was improved. Residents in the Far Cotton area would be given two hours' notice of a flood occurring.
2007	A washland was created in the Upton area where 1.2 million cubic metres of water could be stored during times of flood. Major roads in the area such as the A45 are on embankments up to 6 m high. The railway station has been protected by building 4 m floodwalls at Foot Meadow.

Check your understanding — Tested

List the specific information (facts and figures) in the table about how Northampton has been protected from flooding.

Go online for answers — Online

Exam practice

For a named example, explain how a river has been controlled. **(6 marks)**

Answers online — Online

Exam tip

Look back at the mark schemes for questions on case studies on page viii of the introduction.

Chapter 9 Tectonic Landscapes

Location and characteristics of tectonic activity

The distribution of earthquakes and volcanoes in the world

Revised

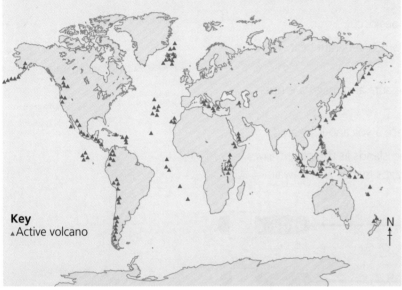

Key
▲ Active volcano

N ↑

↑ **Figure 1** The world's distribution of active volcanoes

Key terms

Location – the place where something can be found.

Characteristics – the main features of the item.

Distribution – the spread of something across the Earth.

Exam tip

You may be asked to describe a distribution on a map.

- You should start with **general** points.
- Your answer should then become more **specific**.
- If data or evidence is asked for, you will lose a mark if you do not include it.
- If data or evidence is not requested, you should still include some as you will be given credit.

Exam tip

If you are asked to **describe**, be careful not to explain as you will not gain any extra marks.

Check your understanding ————————— Tested

Write down five bullet points about the distribution of active volcanoes shown in Figure 1.

Go online for answers ————————— Online

Exam practice

Describe the distribution of volcanoes shown in Figure 1. Use evidence from Figure 1 in your answer.   **(4 marks)**

Answers online

Online

Reasons why earthquakes and volcanoes both occur where they do

Revised

What is plate tectonics?

- Plate tectonics is a theory which gives an explanation for the location of earthquakes and volcanoes.
- The Earth's crust is divided into plates.
- These plates move a few centimetres every year as they are floating on the **mantle**.
- This movement is caused by **convection currents** in the mantle.
- This movement causes the plates to collide or move apart.
- The pressure created at these boundaries is what causes earthquakes and volcanoes.

What are hotspots?

- A hotspot is another place where volcanoes occur.
- Hotspots are fixed points in the mantle which generate heat. They usually occur under oceans.
- The intense heat causes a build-up of pressure and magma erupts through the crust.
- If the magma rises above the ocean surface a volcano is created.
- Hotspots usually create chains of volcanic islands as the plate moves over the hotspot, for example the Galapagos Islands and Hawaii.

> **Key terms**
>
> **Mantle** – the layer of the Earth beneath the crust.
>
> **Convection current** – a movement of heated material up through the mantle from the Earth's core to the crust.

> **Check your understanding**   Tested
>
> How are volcanoes created at hotspots?

> **Go online for answers**   Online

The characteristic features of plate boundaries

Revised

Plate boundaries are the points at which the Earth's plates meet. There are different types of plate boundaries depending on the way that the plates are moving.

> **Exam practice**
>
> Why do earthquakes and volcanoes occur at plate boundaries?   **(4 marks)**
>
> **Answers online**   Online

> **Key terms**
>
> **Convergent (destructive) plate boundary** – the plates move towards each other and destroy the crust.
>
> **Divergent (constructive) plate boundary** – the plates move away from each other and create the crust.
>
> **Conservative plate boundary** – the plates move alongside each other; crust is not made or destroyed.

Convergent plate boundaries

There are three different types of **convergent plate boundary**.

Exam tip
Learn how to draw these simple cross-section diagrams and their labels. You could be asked to draw them or label them in the exam.

1 Where a continental plate and an oceanic plate meet

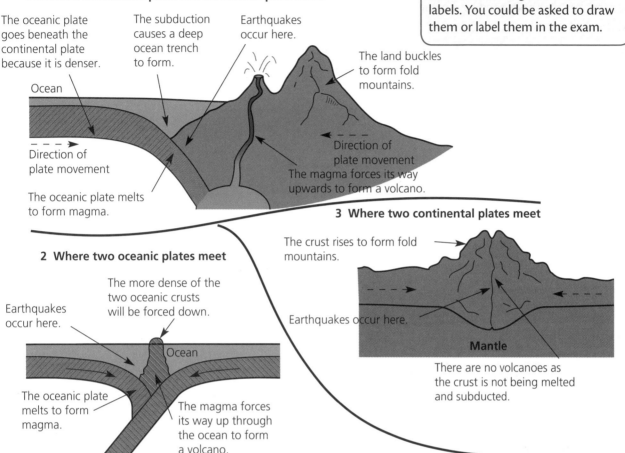

The oceanic plate goes beneath the continental plate because it is denser.

The subduction causes a deep ocean trench to form.

Earthquakes occur here.

The land buckles to form fold mountains.

Ocean

Direction of plate movement

The oceanic plate melts to form magma.

Direction of plate movement
The magma forces its way upwards to form a volcano.

2 Where two oceanic plates meet

The more dense of the two oceanic crusts will be forced down.

Earthquakes occur here.

Ocean

The oceanic plate melts to form magma.

The magma forces its way up through the ocean to form a volcano.

Mantle

3 Where two continental plates meet

The crust rises to form fold mountains.

Earthquakes occur here.

Mantle

There are no volcanoes as the crust is not being melted and subducted.

Divergent plate boundary

These usually form under the ocean. The volcanoes that are created form ocean ridges.

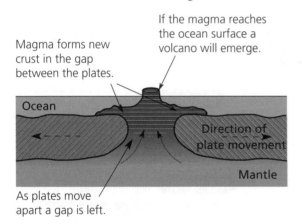

If the magma reaches the ocean surface a volcano will emerge.

Magma forms new crust in the gap between the plates.

Ocean

Direction of plate movement

Mantle

As plates move apart a gap is left.

Conservative plate boundary

This is when plates move alongside each other either in the same direction at different speeds or in different directions.

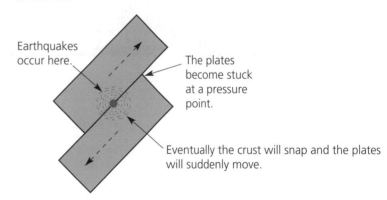

Earthquakes occur here.

The plates become stuck at a pressure point.

Eventually the crust will snap and the plates will suddenly move.

Exam practice

Draw an annotated diagram of a constructive plate boundary.   (4 marks)

Answers online

Online

The methods used to measure earthquake magnitude

Revised

There are two methods used to measure earthquakes: the Mercalli scale and the Richter scale.

The main characteristics of the **Mercalli scale** are the following:

- It is a measurement of the intensity of the earthquake completed from eyewitness accounts.
- It is measured on a scale of 1–12.
- The scale is quite subjective and uses whole numbers.
- The person measuring has to be present at the earthquake site.

The main characteristics of the **Richter scale** are the following:

- It is a measurement of the magnitude of the earthquake completed by a seismograph.
- It is measured on a scale of 1–9 where each number of the scale is split into 10 points.
- The scale is objective.
- The measurement can be done from a different location from the earthquake site.

Check your understanding

Tested

Draw a table to summarise the differences between the two scales.

Go online for answers

Online

Exam practice

State two differences between the Mercalli and Richter scales. **(2 marks)**

Answers online

Online

The characteristics of the focus and epicentre of an earthquake

Revised

Epicentre	Focus
This is on the Earth's surface.	This is a point below the surface of the Earth.
It is the place where the worst effects of the earthquake are felt.	The depth below the surface varies with different earthquakes.
If the area is inhabited, this will be the place where the worst destruction to buildings and **infrastructure** will occur.	This is the point where the pressure build-up is released; **shock waves** move out from this point.

Exam practice

State one difference between the epicentre and focus of an earthquake. **(1 mark)**

Answers online

Online

Exam tip

Remember – **e** is for **e**picentre and **E**arth's surface.

Key terms

Epicentre – a point on the Earth's surface.

Focus – a point below the Earth's surface.

Shock wave – a wave released from the point where an earthquake originates below the Earth's surface.

Infrastructure – the services that are in an area such as roads and electricity.

Management of the effects of tectonic activity

The reasons why people continue to live in areas of volcanic activity

Revised

Topic	Reason	Example
Economic	Volcanic soils are very fertile. Cheap **geothermal energy** is produced from areas which have volcanic activity.	Coffee is grown on the slopes of volcanoes in Colombia. In Iceland, Reykjavik has cheap electricity produced from geothermal energy.
Social	Perception – people think that there is little risk of a volcano happening, or that the government will help them if a volcano does erupt. Family and friends – they have lived there for many years and all their relations and friends are close by. Poverty in LICs – many people in low income countries cannot afford to move.	In the USA when Mt St Helens erupted, people refused to move from the evacuation area. They did not believe that the volcano would erupt because it had been dormant for years. In Sicily where Mt Etna is erupting many families live on the slopes of the volcano. People live on Mt Merapi in Indonesia.
Environmental	The area around the volcano is very scenic and many tourists visit.	Mt Etna in Sicily has many tourist guides so provides jobs for the local people.

Key terms

Geothermal energy – the energy stored within the earth which can be used to produce electricity.

Economic reasons – to do with wealth.

Social reasons – to do with people.

Environmental reasons – to do with the natural and human landscape.

Earthquake-proof buildings – buildings that are built with such extras as flexible frames so that they can withstand earthquakes.

The reasons why people continue to live in areas of earthquake activity

Revised

Topic	Reason	Example
Economic	Many earthquake areas are very scenic; locals get jobs in the tourist industry. People have jobs in the mining industry in earthquake areas.	Many jobs in Iceland are related to showing tourists the earthquake areas. Copper is mined in Chile. In 2007, two people were killed in a mine during an earthquake. However, the local people think the risk is acceptable because they have jobs.
Social	Perception – people think that there is little risk of an earthquake happening (in LICs), or that the government has taken precautions against the likelihood of an earthquake (in HICs).	In Japan many buildings are **earthquake proof** therefore the local people feel safe.
Environmental	Some earthquake areas have been developed for the wealthy and therefore people will wish to live there because of the fantastic environment.	Malibu, a very expensive coastal resort, has developed on the Californian coast of the USA.

Prediction and prevention of the effects of volcanic eruptions and earthquakes

Revised

	Earthquakes	Volcanoes
Prediction (forecasting)	Seismometers are machines that can measure movement. This can predict when an earthquake is likely to take place.	The eruption at Mt St Helens was predicted due to the movement of tiltmeters which had been placed on its slopes. Gases are monitored, for example the eruption of Mt Pinatubo was predicted due to the change in the gases coming out of the volcano. Satellites can be used to record the changes in the shape of a volcano.
Building design and defence	Many buildings now have special features to make them more earthquake proof such as automatic shutters that come down over the windows, and an interlocking framework. For example, San Francisco airport has columns standing on 1.5 m ball bearings. This allows the building to move with an earthquake.	If lava flows are cooled with water they will move more slowly and eventually stop. Diversion channels can also be built around villages. On Mt Etna earth walls have been built to divert lava away from a cable car station which is used by thousands of tourists every year.
Education and planning	In Japan, schools and businesses practise earthquake drills such as ducking under tables. In the USA the government plans for earthquakes by having information packs on how to prepare for an earthquake.	People who live close to volcanoes are taught to look for the signs that may mean that the volcano is likely to erupt. On Sicily the locals are taught to look for changes in the consistency of the lava.

The causes and effects of an earthquake in a named location – case study: Turkey 1999

Revised

Exam tip

You only have to learn one of these case studies so learn *either* Turkey *or* Montserrat.

Turkey earthquake 1999

Causes
- Turkey lies between three continental plates: the Eurasian, the African and the Arabian.
- The movement of these plates caused the earthquake to occur.
- The epicentre of the earthquake was the town of Izmit which is on the North Anatolian fault.
- During the earthquake, the north Anatolian fault moved approximately 3 m.
- Izmit is built on soft rock made from clay and sand.

Effects on people
- Number of people killed – aproximately 18,000 people were killed, many due to buildings collapsing.
- Number of people homeless – approximately 300,000 people became homeless.
- Transport routes destroyed – between Ankara (the capital) and Istanbul (the largest city), the motorway was destroyed which caused disruption to emergency services.
- Psychological problems – people were very distressed.
- Rebuilding costs – these were estimated to be in the region of $10 billion.

Effects on environment
Damage from pollution:
- A fire caused by the earthquake at the Tupras oil refinery meant that 700,000 tonnes of oil were lost.
- The earthquake caused damage to the sewage works at Pekim causing the sewage to leak into local rivers and groundwater.
- There was a leak from the chlorine factory at Yalova.
- The land was raised out of the sea on the coast near the Sea of Marmara.

Check your understanding Tested

1 Produce two tables with the headings 'causes' and 'effects'.

2 In one of the tables write all the specific facts and figures for the Turkey earthquake of 1999. In the other table write all the specific facts and figures for the Montserrat eruption of 1997.

Go online for answers Online

Key terms

Causes of an earthquake – what makes an earthquake happen.

Effects of an earthquake – the impact of the earthquake on the people who live in the area and the built and natural landscape.

The causes and effects of a volcanic eruption in a named location – case study: Montserrat, 1997

Revised ☐

Montserrat volcanic eruption 1997

Causes

- Montserrat is a volcanic island. It is on a destructive plate boundary. The two plates that meet at this boundary are the North American plate and the Caribbean plate.
- In June 1997, Chances Peak volcano erupted.
- It produced a huge ash cloud which covered the southern part of the island. Pyroclastic flows of hot rock and ash were also thrown out of the volcano.

Effects on people

- Number of people killed – 19 people were killed by the pyroclastic flows.
- Villages destroyed – the villages of Farm and Trant were completely buried by the ash flow.
- Number of homes destroyed – approximately 150 houses were destroyed.
- Transport routes destroyed – Bramble airport was closed which disrupted the aid which was coming in to help the islanders. Roads on the island were covered with ash and rocks.
- Psychological problems – people left the island to go to safety and never returned because they had lost everything.

Effects on environment

- Ash and rock covered 4 square kilometres of land.
- Pyroclastic flows caused the River Belham to flood.
- Pyroclastic flows removed all vegetation in the south of the island including the ridges surrounding Farrell's Yard.

Exam tip

Some of the reasons shown on page 61 could be used in more than one category. For example, earthquake-proof buildings are a **social reason** because they make people feel safe, and an **environmental reason** as buildings are part of the human environment.

Check your understanding

Tested ☐

List some more of the reasons which could fit into an exam answer on either earthquakes or volcanoes.

Go online for answers

Online ☐

Exam tips

- In your answers about Montserrat you could include an annotated diagram of a destructive plate boundary like the one on page 59.
- You may have other information you have learnt about the Turkey earthquake or the Montserrat eruption. It is fine to have different information as long as it is correct! Just add it to the table of specific facts that you create in the activity.

Exam practice

Using examples, give reasons why people continue to live in areas of volcanic activity.   **(4 marks)**

Answers online

Online ☐

Exam tip

Look back at the mark schemes for case study questions on page viii of the introduction.

Exam practice

For a named example, explain the causes and effects of a volcanic eruption or an earthquake.   **(6 marks)**

Answers online

Online ☐

Chapter 10 A Wasteful World

Types of waste and its production

Section B – Remember that you have only been taught one of these topics. If you have been taught A Watery World, turn to page 75.

What types of waste are there?

Revised ☐

Some of the different types of waste are:

● **biodegradable** and non-biodegradable

● **domestic** and industrial

● **hazardous** or non-hazardous

● solid or liquid.

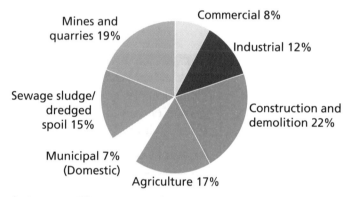

↑ **Figure 1 Different types of waste produced in the UK.**

Mines and quarries 19%
Commercial 8%
Industrial 12%
Sewage sludge/ dredged spoil 15%
Construction and demolition 22%
Municipal 7% (Domestic)
Agriculture 17%

Key terms

Biodegradable – can be broken down by bacteria.

Domestic – waste that is produced by the average household.

Hazardous – dangerous to humans.

HIC – High Income Country.

LIC – Low Income Country.

E-waste – waste such as mobile phones which are discarded before they are broken.

White goods – goods such as fridges and washing machines.

Exam practice

Describe the different types of waste shown in Figure 1. **(4 marks)**

Answers online

Online ☐

What are the differences between HIC and LIC waste production?

Revised ☐

HICs contain 20 per cent of the world's population but consume 86 per cent of the world's products. People in HICs have more money and buy more products. This is known as the consumer society. If people have more things they have more to throw away hence more waste!

Domestic waste in HICs comes from packaging, nappies, **E-waste** and **white goods**.

Check your understanding

Tested ☐

Study the information about the waste produced by HICs. State why LICs do not produce as much waste as HICs.

Go online for answers

Online ☐

An example of HIC and LIC consumption differences

The city of Los Angeles, USA, produces approximately 1,250 kg of rubbish per person each year. This is due to the wealth of the people who live there who consume vast amounts of products and therefore have a lot of waste.

The people in Abidjan, on the Ivory Coast, Africa, only generate 200 kg of rubbish a year! These are poor people without the means to buy many consumer products.

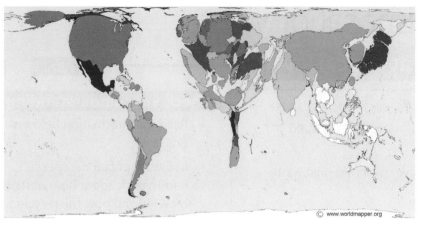

↑ Figure 2 Choropleth map showing waste in selected countries

Exam tip

You may be asked to describe a distribution on a map.

● You should start with **general** points.

● Your answer should then become more **specific** naming countries or continents.

● If data/evidence is asked for, you may lose a mark if you do not include it.

● If data/evidence is not requested, you should still include some as you will be given credit.

Check your understanding ———————————— Tested

Describe the distribution of waste production in Figure 2. What is the level of development of the countries that produce the most waste?

Go online for answers ———————————— Online

What types of domestic waste are produced by HICs? ——————— Revised

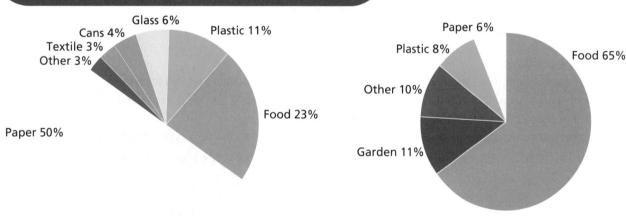

↑ Figure 3 Types of domestic waste in a HIC

↑ Figure 4 Types of domestic waste in a LIC

Check your understanding ———— Tested

Compare the types of waste produced by the USA (a HIC) and Bangladesh (a LIC).

Go online for answers ———— Online

Exam tips

● When the command word **compare** is used, you should state the similarities between the figures. However, examiners at GCSE will usually credit comments about the differences (contrasts).

● You do not need to know the types of waste produced by a LIC but it might be used as a comparison.

Exam practice

Why would the waste produced by Bangladesh (a LIC) be different to that produced by the UK?   **(4 marks)**

Answers online

——————— Online

Recycling and disposal of waste

How is waste recycled at a local level?

Revised

Bracknell Forest Council in Berkshire works with Reading and Wokingham council for the disposal of waste. It has created 're3' which is a waste management partnership.

Bracknell, Wokingham and Reading have two household waste recycling centres:

- Smallmead, Island Road, Reading
- Longshot Lane, Bracknell.

There are also 150 **recycling sites** around the three authorities located at convenient sites such as large supermarkets.

Check your understanding

1. Read the information below on Bracknell.

2. Make a list of all the specific information about how waste is recycled and how the recycled material is used.

Go online for answers

Exam practice

Using a named example, how is waste recycled at a local level? **(4 marks)**

Answers online

Online

How is the recycled material reused?

Revised

Re3, the waste management partnership, have agreements with many different companies to recycle the waste that they collect. Here are how four of the products that are recycled by the residents are turned into usable products.

Key terms

Recycling – making waste products available for re-use.

Recycling sites – places where waste products can be taken to be made available for reprocessing.

Reusing – making waste products into a new product.

Land fill – areas of land that are used to dispose of waste products. Large holes are dug, filled with waste and then covered over with soil.

Incineration – the burning of waste products.

Toxic waste – waste materials that are harmful to humans.

Cans

Cans are baled at a Biffa Waste Management facility in Southampton. They are then transferred to a reprocessing facility in Leicester. They are first split into steel and aluminium and then **reused**, being made into products such as cars or new cans!

How do households in Bracknell recycle their waste?

Bracknell has a two-weekly collection system. One week the following are collected:

- the brown bin – garden waste
- the blue bin – plastic bottles and tins
- the green box – paper.

On the alternate week the green general waste bin is collected. Other types of rubbish such as shoes, foil and e-waste can be taken to the household waste **recycling** centres.

Paper and cardboard

The paper and cardboard are recycled in Maidenhead. The material is sorted and baled. It is then moved to the St Regis paper mill in Kent where it is turned into new packaging material.

Plastics

Plastic bottles are recycled by Baylis Recycling at their plant in Keynsham near Bristol. The bottles are sorted into different types of plastic. Products made from used plastic include garden furniture, fleece jackets and, yes, new plastic bottles.

Glass

The glass is reprocessed in Yorkshire. The glass is washed and crushed and then mixed with raw materials such as sand and limestone to make new glass containers.

How does Germany (a HIC) dispose of its domestic (municipal) waste?

Revised

Landfill
- There are 160 landfill sites.
- Waste has to be treated before it is allowed to be put into landfill.
- One treatment plant is owned by The Group and is sited in Luebeck in Germany. It can treat 200,000 tonnes of domestic waste annually.

Incineration
- There are 68 incinerators in Germany.
- The plant at Darmstadt incinerates 212,000 tonnes of waste a year.
- Other incinerators are mini-power plants which burn waste and provide energy to local homes and businesses.

14 million tonnes of municipal waste

Recycled waste
- Germany recycles 60% of its waste.
- All products that can be recycled have the 'Grune Punkt' emblem. Buying these recyclable products costs the average family between £100 and £200 a year which has been added to their cost by the producer to pay for the recycling. Plastic, glass, paper and cardboard are all recycled. There are a number of problems with the system. The scheme is so successful that Germany cannot recycle everything that is collected. They have had to export their recyclables to other European and Asian countries. For example, plastic shampoo bottles from Oggersheim are turned into sandals in Indonesia.

How does Germany (a HIC) dispose of nuclear waste?

Revised

- At present, low-level waste is stored around the country at about 50 locations.
- German high-level waste is stored in Siberia and recycled in the UK and France.
- By 2013 Germany will store all of its low-level waste, 95 per cent of the total nuclear waste, at Konrad.
- By 2025 Germany will also have a facility at Gorleben to deal with the remaining 5 per cent of high-level waste.

Exam practice

1 Landfill is the best way to dispose of waste. Discuss this statement. **(3 marks)**

2 Using a named example, explain how an HIC disposes of its waste. **(6 marks)**

Answers online

Online

Check your understanding

Tested

In your opinion, which is the best way to dispose of waste? Give reasons for your answer.

Go online for answers

Online

How does Germany (a HIC) dispose of other toxic waste products?

Revised

"Don't tell me where you put it, just so long as you get rid of it."

Germany exports its **toxic waste** products. Between 1990 and 1995, shiploads of German toxic waste were sent to Poland, Estonia, Egypt and Albania.

During 1991 and 1992, 480 tonnes of toxic waste (pesticides) were sent to Albania. In Germany, it would have cost the company $5500 per tonne to dispose of the pesticides. The pesticides were sent to Albania to be used by farmers even though they were too dangerous to use in Germany.

However, within the shipment were 6000 litres of toxaphone, which is highly toxic and can kill wildlife. The Albanian government did accept the shipment but when they realised what it was they asked the Germans to take it back. Eventually other world nations forced Germany to take it back. One part of the shipment was found at Bajza in Albania, where the pesticides were seeping into Lake Shkodra.

The advantages and disadvantages of the different ways that Germany disposes of its waste

Revised

Types of waste disposal	Advantages	Disadvantages
Landfill	It uses up land such as former quarries which can be filled in and the land reused. It is a cheap option.	Land is used up to dispose of waste. The waste can rot and cause problems for later generations.
Incineration	The heat that is produced can be used to generate electricity. Jobs are created.	Fumes are omitted when the waste is burnt. The plants are visually polluting to the surrounding area.
Recycling	Less waste is put into landfill sites. Fewer resources are used as waste is being made into something new.	It is very difficult to enforce the law on firms. Germany has a lack of recycling facilities. It has had to export to other countries which is costly.
Export	Other countries have to deal with German waste. This means – for nuclear waste – they do not have dangerous radioactive waste in their country. Less waste is put into landfill sites or has to be incinerated.	It is expensive to pay other countries to deal with your waste. Internationally Germany had a bad reputation due to the way that certain German companies were sending dangerous waste to LICs.

Check your understanding

Tested

1   Read the information on Germany.
2   Make a list of all the specific information about how Germany disposes of its waste.

Go online for answers

Online

Sources and uses of energy

Key terms

Renewable – energy sources that can be reused and therefore will not run out, such as wind and the Sun. They are known as infinite resources.

Non-renewable – energy sources that once they have been used can never be used again, such as coal and oil. They are known as finite resources.

Exploitation – the use of an energy source.

Development – how the resource is then used to produce power.

Greenhouse gas – a gas such as carbon dioxide which is 'allegedly' causing global warming.

Production of a resource – how the resource is gained from the environment.

Visual pollution – something that is displeasing to the eye.

Check your understanding
Tested ☐

What is the difference between **renewable** and **non-renewable** energy sources?

Go online for answers
Online ☐

Attitudes to the exploitation of energy sources
Revised ☐

These will be different dependent upon the type of energy being **exploited**, the time when it is being exploited and the needs of the local inhabitants.

Exam tip
You will need to know the attitudes of people towards one renewable and one non-renewable energy source.

Coal as an energy source

● In the UK, coal was exploited in a time when people needed jobs and were less concerned about their environment.

● Companies now assure local communities that the land will be returned to its previous condition.

● On a global scale, China is basing its **development** on coal as an energy source which is causing concern for the global community because of the **greenhouse gases** that burning coal emits.

Wind as an energy source

● Many local residents in countryside areas object to the building of wind farms on hills close to their settlements because they have not received the benefits such as cheaper energy bills or employment. However, residents and businesses close to the wind turbine in Green Park in Reading have benefited with cheaper electricity bills and therefore do not object to the turbine.

● There are few objections to wind turbines from a global perspective as their environmental impact is less than that of other forms of energy production.

The production and development of one non-renewable energy source – coal

Revised

Production	Development
Coal is found in many countries around the world.	When coal is burnt greenhouse gases are emitted.
Coal is cheap to mine because it is just below the surface and can be mined very easily, for example, it is mined in Australia.	Acid rain is produced when coal is burnt. This has caused problems in the forests of Scandinavia where roots and leaves have been destroyed.
Waste heaps are left close to coal mines.	It is relatively easy to convert coal into energy by simply burning it.
Deep shaft mining can be dangerous, for example, nine miners died in China in May 2012 when a shaft collapsed.	Coal supplies should last for another 250 years.

The production and development of one renewable energy source – wind

Revised

Production	Development
Some greenhouses gases are given off during the production of the turbine and when it is transported to its site.	New wind turbines are quiet and efficient.
Some people think that the turbines create **visual pollution.**	Wind turbines do not give off greenhouse gases.
The wind is free.	Wind turbines can be on land or at sea.
Turbines are relatively cheap, costing £1500 for a 1kW wind turbine.	Turbines can be unsightly/visually intrusive.
Offshore turbines may disturb migration patterns of birds.	There needs to be an annual local wind speed of more than 6 metres per second.

Exam practice

Compare the advantages and disadvantages of producing different types of energy. **(4 marks)**

Answers online

Online

The global energy mix of energy consumption

Energy consumption varies greatly around the world. This is due to a number of different factors which include population, income and wealth, and the availability of energy supplies.

Factor	Effect
Population	Energy consumption has increased because: • between 1990 and 2008 the world population increased by 27 per cent • the average use of energy per person has globally increased by 10 per cent. This is not equally spread – the countries with higher populations do not necessarily have the highest consumption of energy.
Wealth and income	Countries that are wealthier will use more energy than poorer countries. If a country is wealthy it will be able to provide the energy required by its population. The population will earn enough money to enable them to buy electrical equipment.
Availability of energy supplies	Countries that have a plentiful supply of energy may have high consumption. However, the rate of consumption also depends on other factors such as the wealth and income of the population.

Country	Total energy consumption per capita per annum (2003) [kgoe/a] kilogrammes of oil equivalent	Total population (2005) (millions)
Bangladesh	161	141
Brazil	1,067	186.4
Qatar	21,395	1
UK	3,918	60
Venezuela	2,057	26

Exam tip

You will need to be able to explain figures such as the ones in the table on the left.

Check your understanding

1 Qatar and Venezuela both have a plentiful supply of energy. Why is their consumption so different?

2 The UK has a smaller population than Bangladesh but its energy consumption is much higher. Why?

Go online for answers

Online

Exam tip

You will need to have an awareness of the energy consumption of different countries in the world and be able to explain it.

The exploitation of energy resources has a varied impact on the environment

Revised

The ash that is released when coal is burnt has been found to be radioactive. People who live within 1 mile of a power station may be ingesting small amounts of radiation, as will crops grown around the power plants.

Waste heaps are produced. The heaps cause visual pollution and are unstable. For example, in 1966 a waste heap in Aberfan in Wales collapsed killing 116 children.

Non-renewable energy source – coal

Coal burning to produce electricity also contributes to global warming as carbon dioxide is released as a by-product.

Acid rain is formed from sulphur dioxide and nitrogen oxide which are released when coal is burnt. They make the rain that falls more acidic. This can have an effect on plants and animals. Coal-burning power stations in northern Europe are responsible for acid rain in southern Scandinavia. In this way, Britain has contributed at least 16 per cent of the acid rain that has fallen on Norway.

Exam tip
You need to know two case studies: one for a non-renewable energy source and one for a renewable energy source.

Wind turbines do not give off carbon dioxide when they produce energy. However, the production of the turbine, its transportation to site, maintenance and disposal at the end of its life does produce carbon dioxide.

Wind farms take up a large area of land. However, other forms of land use can take place in the same area, for example farming can take place on fields which are also used for wind turbines.

Renewable energy source – wind

Wind farms affect the wildlife of an area. Offshore wind farms can affect the migration patterns of birds and wind turbines do kill a number of birds every year. In Denmark, where wind turbines generate 9 per cent of the electricity, 30,000 birds are killed per year. However, new turbines are more bird friendly.

Wind turbines also have an impact on bats. Large numbers of bats were being killed by wind turbines. In the eastern USA, wind turbines are now switched off when wind energy is low as this is when bats are most active. This has led to a 73 per cent drop in bat deaths.

Management of energy usage and waste

How energy is being wasted – the domestic situation

Revised

Energy is lost from a home 25 per cent through the roof, 35 per cent through walls, 15 per cent through the floor, 10 per cent through windows and 15 per cent through draughts.

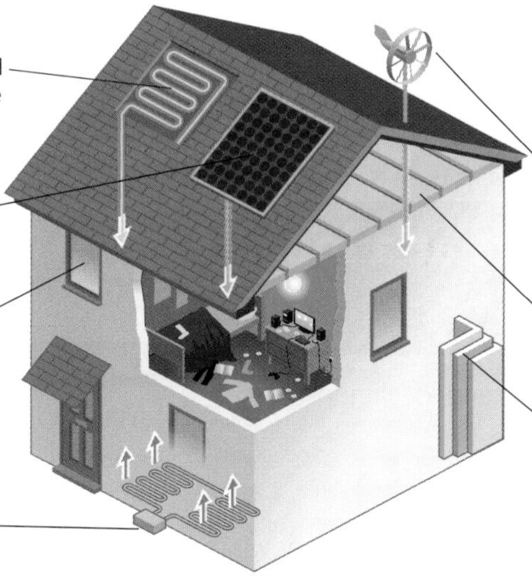

Solar water heating. The fluid in the panel is warmed by the sun. This produces hot water for the house.

Photovoltaic electricity. The panels produce electricity from light.

Double glazing. Two panes of glass insulate house.

Ground source heat pump. Takes heat from the ground and uses it to heat the house.

← Electrical power into house
← Heat into house
← Heat to water supply in house

Micro wind turbine. The force of the wind produces energy which is stored in a battery.

Loft insulation. Stops heat escaping through the roof.

Cavity wall insulation. Stops heat escaping through the walls.

↑ Figure 5 How energy can be saved in the home

Check your understanding
Tested

How could energy be saved in the teenager's bedroom?

Go online for answers
Online

How energy is being wasted – in industry

Revised

These are some of the ways that British industry wastes energy.

● Office devices are left running, for example, computers and printers. Over the Christmas period, the amount of energy wasted over the ten-day Christmas shutdown is enough to roast 4.4 million turkeys.

● Approximately £1 in every £12 spent on fuel is being wasted by the UK steel industry.

Countries have carbon footprints of differing sizes

Revised

A carbon footprint is a measure of the impact that human activities have on the environment in terms of the number of greenhouse gases they produce.

● The least developed, low income countries will have the lowest carbon footprints.

● The most developed, high income countries have the highest carbon footprints.

Exam tip
You will need to be able to look at statistics and work out what they show about the carbon footprint of different countries.

Exam practice

1 What is meant by the term carbon footprint? **(1 mark)**

2 Describe three ways in which energy is wasted in the domestic situation. **(3 marks)**

Answers online

Online

Solutions to energy wastage at a national scale

These are some of the ways that the government is trying to reduce energy wastage.

- A grant of £2500 is available per household for green technologies from the Low Carbon Buildings Programme.
- Planning permission is no longer required to install wind turbines, solar panels, ground and water source heat pumps and biomass systems.
- There are grants of up to £1 million for public buildings such as schools and hospitals where the government will pay up to 50 per cent of the start-up costs of renewable energy. For example, Howe Dell Primary School in Hatfield now has a wind turbine and solar panels, and recycles its rainwater.

Exam tips

- In an exam question, if you are asked for examples referring to solutions to energy wastage, ensure that you write about named places and specific information.
- Make sure that you write about the scale that is being asked about in the question – national or local.

Solutions to energy wastage at a local and domestic scale

Many local councils are encouraging home owners and local industry to use renewable energy.

Wind power

Over the past five years there has been an increase in the number of wind turbines which can be seen around the UK. Large numbers of them feed into the National Grid. Others provide for more local electricity demands. The turbine at Green Park in Reading has been providing energy since 2005 for 1500 homes and businesses.

Reduction of council tax

British Gas is working with sixteen local councils, including Reading Borough Council, to improve energy efficiency. If households implement energy efficiency measures such as loft insulation, they will receive £100 off their council tax bill.

Examples of how some local councils are encouraging residents to save energy

- **Oldham City Council** Oldham is helping its residents to save energy by upgrading their social priority housing, for example installing loft insulation and giving ideas on their website on how to save energy such as 'using one energy-saving bulb can save £5 a year in energy costs'.

- **Penwith Housing Association, Penzance, Cornwall** In Cornwall ground source heat pumps have been fitted to fourteen bungalows by one local council. These provide the householders with heating via radiators and hot water. The cost of the project was £200,000, much of which was obtained through grants. Due to the success of this scheme there are now 700 schemes across the country which are either running or being installed.

Check your understanding

Make a list of the different councils and what they are doing to save energy.

Go online for answers

Exam practice

Explain solutions to energy wastage on a local and domestic scale. **(4 marks)**

Answers online

Online

Exam tip

Look back at the mark schemes for case study questions on page viii of the introduction.

- **Woking Borough Council** Woking Borough Council has its own electricity wires and therefore does not have to pay to be part of the National Grid. A CHP system has been installed which powers six buildings in the town centre and some sheltered housing close by. A new power plant at Woking Leisure Centre supplies the pool and 136 homes close by.

By 2011 the council wants to provide 20 per cent of energy through renewable resources. It has more photovoltaic cells than anywhere else in the country, for example there are photovoltaic cells at Vyne Community Centre.

Chapter 11 A Watery World

Water consumption and sources

Section B – Remember that you have only been taught one of these topics. If you have been taught A Wasteful World, turn to page 64.

Key terms

Water consumption – the amount of water used.

Domestic usage – water used by households.

Showering society – HICs, where many people shower every day.

Agricultural usage – water used by farmers.

Irrigation – the artificial watering of the land.

Industrial usage – water used by factories.

Cottage industries – small-scale production usually completed in a domestic environment.

Exam tip

You could be asked what the differences are in water consumption **generally** between HICs and LICs, or you could be asked more **specifically** about the differences between domestic usage by HICs and LICs.

The differences in water consumption between LICs and HICs

Revised ☐

- The usage of water in HICs is much greater than that in LICs, for example people in the USA use 600 litres per day and people in Ethiopia use 25 litres per day.
- The way water usage is divided between domestic, agricultural and industrial use is dependent upon the country being discussed, but as a general rule HICs use the majority of their water domestically and in LICs the highest usage is for agriculture.

Exam practice

State two ways in which water usage differs between HICs and LICs. **(2 marks)**

Answers online

Online ☐

Reasons for these differences

Revised ☐

Domestic usage

- In HICs, people have water piped into their homes. This allows them to use water in washing machines and dishwashers.
- HICs are also known as a '**showering society**' because many people shower every day.
- All of these activities create a greater demand for water in HICs than LICs.
- In LICs many people do not have piped water to their homes so wash their clothes and dishes in local streams. The same water is used for many different purposes.

Agricultural usage

- In HICs, **irrigation** systems are used which are run by computers. The computer can determine exactly how much water is required and supply the water quickly, up to 75 litres per second.
- In LICs, plants are watered using buckets or very simple irrigation systems which supply water slowly at approximately 1 litre per second.

Industrial usage

- In HICs, factories are on a large scale and use thousands of litres of water.
- In LICs small-scale **cottage industries** use only a small amount of water.

Greater wealth and increasing levels of development have contributed to increasing water consumption

Revised

- The increase in wealth in HICs has seen a growth in the use of labour-saving machines such as washing machines and, in the last twenty years, dishwashers because people can afford to buy them. These machines use much more water than traditional ways of washing clothes and dishes.

- With greater wealth, people can afford to equip their homes with more bath and shower rooms, enabling people to wash more frequently.

- The increase in wealth has been mirrored by a growth in the leisure and tourism industry. As people demand more facilities, such as golf courses and swimming pools, there will be a corresponding increase in demand for water.

Exam practice

How has greater wealth contributed to an increase in water usage in HICs?

(3 marks)

Answers online

Online

On a local scale our water comes from groundwater, reservoirs and rivers

Revised

Groundwater

Some areas of the UK get their water directly from the ground. This is the case in the Thames Valley. The rocks below the Thames Basin are in layers. Chalk, which is **porous**, lies above an **impermeable** layer of clay. This traps water in the layer of chalk forming an **aquifer**. The level of the water in the rock is known as the **water table**. Water is extracted from the aquifer by drilling down into the rock and the water either comes to the surface naturally or is pumped. It is then stored in reservoirs until it is needed.

Reservoirs and rivers

Reservoirs store water and supply it when it is needed. For example, in the west of the Thames Valley, there are ten **raised storage reservoirs**. This includes the Queen Mother Reservoir which is 1 km in diameter, making it the largest area of water in the south of England. The reservoirs are supplied with water drawn from the Thames at times of high river flow. When needed the water is piped to four water treatment works and then to homes and industry.

Other reservoirs are built in the valley of a river, for example Kielder Water on the North Tyne River. The reservoir can hold 200 billion litres of water which it supplies to cities such as Newcastle. The reservoir holds back the river water and releases it on demand. The water is transferred through pipelines to other rivers in the area such as the Tees to enable it to supply water to Middlesbrough.

Key terms

Porous rock – rocks that soak water up like a sponge.

Impermeable rock – rock that does not allow water through it.

Aquifer – underground store of water in permeable rock.

Water table – the level of water within a rock.

Raised storage reservoirs – large bodies of water that are stored above the level of the ground.

Exam practice

Describe one way that we obtain water on a local scale.

(2 marks)

Answers online

Online

Check your understanding

Tested

State three ways in which we obtain water on a local scale. Give an example for each.

Go online for answers

Online

Water surplus and deficit on a world scale

Revised

Areas of the world that have a surplus of water are usually those which receive the most rainfall. This is shown by looking at Figures 1 and 2. There is also a link between the demand for water in an area and the supply, for example the east coast of the USA receives approximately 1000 mm of rain a year but is **water neutral**, not surplus, because of the large population in the area.

Check your understanding
Tested

Research and learn the names of countries which have:

- high rainfall
- water surplus
- low rainfall
- water deficit.

These may be the same countries or they may be different ones.

Go online for answers
Online

Exam tip

Exam questions are unlikely to expect you to remember areas of the world but you should know the continents and the names of countries that have extremes of rainfall and water supply.

Key terms

Water surplus – an area has more water than it needs to supply its population.

Water deficit – an area does not have enough water to supply its population.

Water neutral – an area has the right amount of water to supply its population.

Check your understanding

Describe the distribution of water supply shown in Figure 2.

Go online for answers

Exam tip

You may be asked to describe a distribution on a map.

- You should start with **general** points.
- Your answer should then become more **specific**.
- If data is asked for, you will lose a mark if you do not include it.
- If data is not requested, you should still include some as you will be given credit.

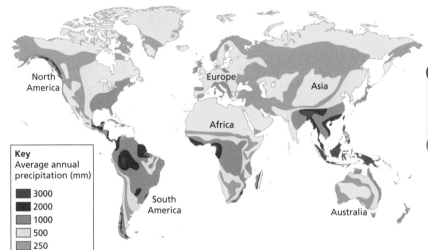

Key
Average annual precipitation (mm)

- 3000
- 2000
- 1000
- 500
- 250
- Below 250

↑ **Figure 1** Map of the world showing average annual precipitation

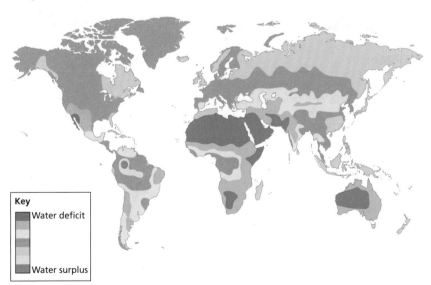

Key
Water deficit
Water surplus

↑ **Figure 2** Map of the world showing areas of water surplus and deficit

Water supply problems

Water supply problems in HICs

Revised

Problem	Description of the problem
Availability	In some areas of the world water is more available than others, even in HICs. This is due to the lack of rainfall but also because of the demands on water because of large populations. For example, most of Australia has a water deficit, although some parts of Australia have plenty of rainfall.
Quality	The quality of water in HICs is continually monitored by the water providers to ensure it is of a good standard. At times water courses can become polluted by factories or agricultural run-off. However, usually the water is of a high quality.
Spatial variability	In many HICs there is a problem of spatial variability in their water supply. This means that the majority of the rain falls on one area but the population lives in another area of the country. For example, in the UK 33 per cent of the population live in the south-east of the country which receives annual rainfall of 625 mm. However, 2000 mm of rain falls on scarcely populated areas of Wales and Scotland.
Seasonal variability	This means the difference in rainfall at different times of the year. For example, countries such as Greece receive approximately 12 per cent of its rainfall in the summer months. This is made worse by the increase in demand caused by the thousands of tourists who go to the area at that time of year.
Loss through broken pipes	The UK's domestic water supply is carried through pipes which are approximately 150 years old. Many of these pipes are cracked and leaks occur. It is estimated that 30 per cent of the water supplied is lost.

Exam practice

Explain two water supply problems in HICs.   (4 marks)

Answers online

Online

Exam tip

You could be asked questions about general water supply problems in HICs or more specific problems, for example, about the loss of water through broken pipes.

Water supply problems in LICs

Revised

Problem	Description of the problem
Clean piped water	In HICs it is taken for granted that people will have piped water to their homes. However, approximately 1 billion people in LICs do not have safe water to drink.
Water-borne disease	These are diseases such as dysentery, which are caused by drinking dirty water. Other water-borne diseases include bilharzia, which is caused by the larvae of snails which get into people's blood streams and cause kidney failure.
Pollution	This can be related to resource exploitation. In the Amazon rainforest in Ecuador, people wash and drink water that is contaminated by oil toxins from unlined pits left by Texaco after they had drilled for oil in the area. This has caused increased rates of cancer in the area.

The management of water usage and resources

The management of water usage in HICs

Revised

Many homes have been fitted with water meters. People with meters use less water as they are charged for how much water they use rather than a fixed amount for the year.

Short-flush toilets – new toilets are fitted with a short flush option which means that less water is used to flush the toilet.

New irrigation systems have been developed. Drip irrigation gives water straight to the root of the plant rather than spraying it in the leaves. This means that no water is lost through evaporation.

If there is a drought, water companies will issue hosepipe bans. If people are seen using a hosepipe they will be fined.

The techniques used in domestic, industrial and agricultural contexts to manage the usage of water

Rain sensors are devices activated by rainfall. They are connected to an automatic irrigation system which then shuts down if it starts to rain.

Walkers Crisps have reduced their usage of water in their production plants by 50 per cent. This has been achieved by:
- installing 30 water meters in their plants
- recycling water in their production process
- educating their workforce about the use of water.

Other companies have used the following ways to save water:
- tap restrictors, which only allow a certain amount of water per usage of the tap
- push button taps so that taps are not left running
- push button showers in washrooms and leisure complexes.

Exam practice

Outline how water is managed domestically in HICs.   **(3 marks)**

Answers online

Online

The management of water usage in LICs

Revised

Recycling systems
In India, the population is growing rapidly. They have decided to recycle water to improve supply problems. Kolkata, which is the capital of the state of West Bengal, recycles its sewage water into clean drinking water. The processing plants have been built by the company Unitech Water Technologies Ltd.

The appropriate technology techniques used to supply water in small communities

Water conservation
Rainwater can be collected from roofs. It is stored in tanks before being poured through a simple filtering process and then used as drinking water. In other areas, water is collected in pools and behind dams until it is needed.

Boreholes
These are holes in the ground which lead down to water-bearing rocks. Holes were dug in the Katine county of Uganda in 2008. They are looked after by the local community who call in a mechanic if the pump goes wrong. In this way the community feel in charge of their water supply and know what to do if things go wrong.

The management of water resources

A dispute between countries over water transfers – the Tigris–Euphrates river system

The Tigris and Euphrates Rivers both have their sources (start) in Turkey then flow through Syria and Iraq to the Persian Gulf. This has caused conflict between the three countries because they all require the water for their countries but Turkey controls the flow of the rivers. As Iraq is the last country that the rivers flow through it receives the least water because of the dams that the other countries have built.

A war between Syria and Iraq was narrowly averted in 1975 after the building of both the Keban Dam in Turkey and the Tabaqah Dam in Syria combined with a drought created serious water supply problems for Iraq.

Syria completed the Tabaqah Dam on the Euphrates River in 1973 forming Lake Assad. This has doubled Syria's irrigated land and provides HEP. However, in 1974 Iraq put troops on its border with Syria and threatened to destroy the Al-Thawra Dam on the Euphrates.

In the 1980s Turkey started the Southeastern Anatolia Project (GAP) which involves the building of 21 dams on the rivers in Turkey. This project will restrict the flow of water into Syria and Iraq. The largest dam – the Ataturk – was completed in 2005.

Turkey states that GAP is beneficial to Syria and Iraq because the dams will regulate the flow of the rivers.
It is estimated that after the GAP project is completed the flow of the Euphrates will decrease:
- at the Syrian border with Turkey from 30 billion cubic metres a year (BCM/a) to 16 BCM/a
- at the Iraq border with Syria from 16 BCM/a to 5 BCM/a.

Turkey signed agreements with Iraq in 1984 and Syria in 1987 to guarantee a minimum flow of 500 cubic metres per second. This has not been upheld.

In January and February 1990 Turkey reduced the flow of the Euphrates by 75 per cent while it filled Lake Assad behind the Ataturk dam. Iraq threatened to bomb the dam so Turkey mobilised its army in defence of the dam and threatened to cut off the water flow completely. Luckily the conflict was averted.

Iraq is concerned that due to the uncertain rainfall in the area soon very little water will be flowing into their country. More than 50 per cent of the farmers in the south of the country have left the area and gone to the cities to find work. The size of many farms has been reduced by 75 per cent because of the lack of water for irrigation.

Check your understanding

Tested

Write a list of all the conflicts that are mentioned in the Tigris–Euphrates water conflict.

Go online for answers

Online

Exam tip

To answer a question on this case study you will need to know specific points and be able to explain what the conflict is about.

A water management scheme – the Three Gorges Dam

Revised

This is a dam on the Yangtze River in China.

Reasons for the water management scheme

- To produce electricity from **hydro-electric power** for the rapidly growing industrial areas in China. The dam is capable of producing 22,500 megawatts of electricity.

- To improve navigation along the river. The flow of the river will be greater allowing larger ships to reach cities further inland. Tonnage of river transport is predicted to increase from 10 million to 50 million tonnes a year.

- To control flooding. Because the flow of the river can be regulated it will stop the river flooding and destroying crops.

> **Key term**
> **Hydro-electric power** – electricity produced from water turning turbines.

The effects of the water management scheme

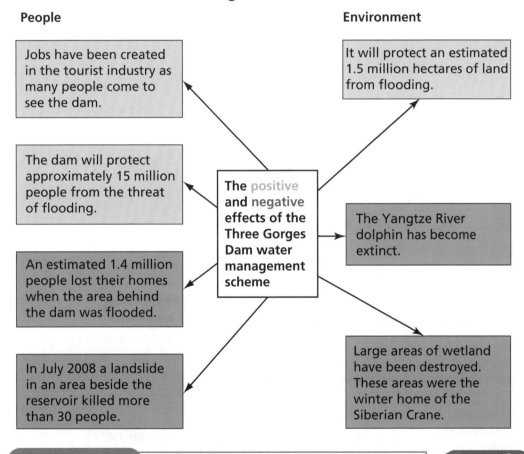

People

Jobs have been created in the tourist industry as many people come to see the dam.

The dam will protect approximately 15 million people from the threat of flooding.

An estimated 1.4 million people lost their homes when the area behind the dam was flooded.

In July 2008 a landslide in an area beside the reservoir killed more than 30 people.

The positive and negative effects of the Three Gorges Dam water management scheme

Environment

It will protect an estimated 1.5 million hectares of land from flooding.

The Yangtze River dolphin has become extinct.

Large areas of wetland have been destroyed. These areas were the winter home of the Siberian Crane.

> **Exam practice**
>
> Explain the effects of the Three Gorges Dam water management scheme. **(6 marks)**
>
> **Answers online**
>
> Online

> **Exam tip**
> Remember to learn the positive and negative effects of the water management scheme for both people and the environment.

> **Exam tip**
> Look back at the mark schemes for case study questions on page viii of the introduction.

Chapter 12 Economic Change

Changes to different economic sectors

The importance of different sectors of employment in countries at different levels of development

Revised

There are three sectors of industry: **primary**, **secondary** and **tertiary**. The importance of these sectors of industry differs with the level of development of the country. The pie charts in Figure 1 show the importance of primary, secondary and tertiary industry in Germany (**HIC**), Taiwan (**MIC**) and Mali (**LIC**).

Exam tip

Learn the definitions of the different sectors and an example of an industry in each one.

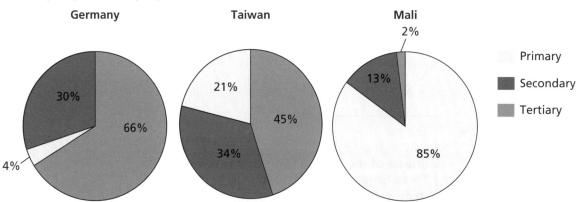

↑ Figure 1 Pie charts showing level of development between Germany, Taiwan and Mali

Key terms

Primary industry – the extraction of raw materials from the ground or the sea. Examples: mining, farming, fishing and forestry.

Secondary industry – the processing (manufacturing) of goods using the raw materials from primary industry. Example: making paper from wood.

Tertiary industry is concerned with doing a service. Examples: teachers, solicitors, sales assistants and cleaners.

High income country (HIC) – rich countries with high GDP such as the USA.

Middle income country (MIC) – countries that are starting to develop and gain wealth such as Brazil.

Low income country (LIC) – poor countries with low GDP such as Ethiopia.

Exam practice

Define the terms primary, secondary and tertiary industry. Give an example of each sector of industry.   **(6 marks)**

Answers online

Online

Check your understanding

Tested

Describe the differences between the employment patterns shown in Figure 1.

Go online for answers

Online

Employment patterns change over time

An example of a country which has seen changes in its employment patterns over time is the UK. These changes are shown in Figure 2. As countries develop, there is a movement away from primary employment into secondary and tertiary.

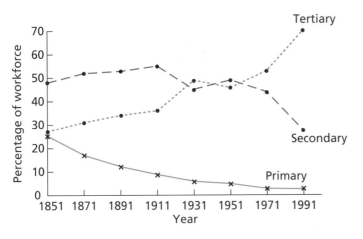

↑ Figure 2 Graph showing changes in UK employment sectors

Exam practice

1 Justify how you would know that Germany is a HIC and Mali is a LIC. Use evidence in your answer.

(4 marks)

2 Give reasons for the changes shown in Figure 2. **(3 marks)**

Answers online

Online

Check your understanding

Tested

Describe the changes shown in Figure 2. Use evidence in your answer.

Go online for answers

Online

Reasons for the decline in the numbers employed in the primary sector in the UK

Depletion of resources
Raw materials that used to be mined in the UK have run out. This has caused a decline in the mining industry.

Cheap imports
Most raw materials now used in the secondary industry in the UK are imported from abroad. It is cheaper to import them because the raw materials that are left in the UK are deep underground.

Reasons for the decline in the primary sector

Social change
- Primary industry jobs such as farming are seen to be hard work.
- Many employees working in primary industries have no chance for promotion.
- The jobs in primary industries are often low paid.

Mechanisation
Fewer workers are needed in primary industries because their jobs are now being done by machines.
- On farms, the jobs that used to be done by farm workers are now done by machinery, for example potato picking.
- Animals are being fed and watered automatically rather than by people, for example chickens are reared in large sheds and computers determine when they are fed by machines.
- Fishing boats now have computers to aid them to find shoals of fish and machinery to work the nets that catch the fish.

Key terms

Raw materials – resources that are used to make useful products.

Mechanisation – the use of machinery instead of people to do jobs.

Government attitudes towards the value of primary industry

The way that governments view the sectors of industry varies between places and over time.

- A country's economy may be based on primary industry such as farming, fishing or mining. This means that the value of primary industry will be high.
- Some countries have lots of primary resources such as oil. In these countries primary industry continues to be of value. This is the case in countries like Saudi Arabia.
- In many HICs in the western world, primary industries such as farming are of value to feed their populations. However the main sector of industry is tertiary.

Exam tip

Remember to learn examples as well as the reasons for the decline in primary and secondary industries.

Check your understanding　　　　　　　　　　　　　　　　　　Tested

Why do government attitudes towards the value of primary industry differ?

Go online for answers　　　　　　　　　　　　　　　　　　　　Online

Reasons for the decline in the secondary sector in the UK

Globalisation
- Modern communication, such as the internet, allows businesses to communicate globally. This means that the owner of a company could be in one country and the goods produced in another.
- Developments in transportation mean that goods can be moved around the world quickly and efficiently.

Mechanisation
The increased use of machinery in secondary industry has led to a decrease in the number of people employed. For example, the car industry used to employ people on all parts of the production line to make the car. Now, in the Honda factory at Swindon, robots are used on some parts of the production line, for example, to paint the cars.

Reasons for the decline in the secondary sector

Cheaper production in LICs and MICs
- Goods can be produced more cheaply in LICs and MICs than in the UK because workers are poorly paid and land is cheap.
- There are also fewer regulations in LICs as governments may not be so worried about environmental impacts, so money does not need to be spent on waste disposal and reprocessing.

Government policies
The withdrawal of government help to industries may have been seen as a cause of the decline of the secondary sector in the UK. In 1967, the British Steel Corporation was formed. This was a nationalised company owned and run by the government to try to protect the production of steel in the country. The company was sold back to private ownership in 1988 when the government's attitude changed towards secondary industry. The UK government tried to stop the decline in secondary industry by offering grants and loans to foreign investors to encourage them to set up their factories in the UK. This was a policy followed in the 1980s and 1990s.

Key terms

Globalisation – the growing economic interdependency of countries worldwide.

Global superhighway – the internet.

Government attitudes towards the value of secondary industry

The way that governments view secondary industry varies from place to place and over time. In HICs the secondary sector has declined and much of the manufacturing is now done abroad in MICs. The government in some countries, for example in the UK, has tried to keep some manufacturing in the country by offering development grants to foreign manufacturers. In the late 1980s Toyota were offered incentives to set up a plant at Burnaston in Derbyshire. The UK government now concentrates on the development of tertiary industry as it sees this as a more valuable sector of industry.

Check your understanding — Tested

1 Which factors contributed to the decline of the primary and secondary industries in the UK?

2 Why do government attitudes towards the value of secondary industry differ?

Go online for answers — Online

Case study – the growth of the secondary industry in China

Reasons for the growth

Factor	Reason for growth
Government policy	All industry was owned by the government but now 20% of companies are privately owned with investors from other countries.
Infrastructure	Many new roads have been built by the government to improve transport in the country.
Energy	The government has built many new nuclear and hydroelectric power stations to power the new industries.
Education	There is a large unskilled workforce. Over the next twenty years approximately 500,000 million people will leave the Chinese countryside in search of work in the cities. There is a growing skilled workforce.
Safety for the workforce	There are few rules for health and safety in Chinese factories which means they can produce goods cheaply.
Wages	People in China will work for less than people in other countries such as the UK.
Raw materials	China has a lot of natural resources such as coal, oil and natural gas.
Trade	China has a long coastline with major ports on trade routes.

Check your understanding — Tested

Which of the reasons in the table are due to human factors and which are due to physical factors?

Go online for answers — Online

Effects of the growth

Some of the effects of the growth have been positive while others have been negative.

The people who live in **urban** areas can now get jobs which give them a higher standard of living.

Many new homes are being built in urban areas which the workers can afford. People are moving out of shanty towns.

760,000 people a year die from illnesses related to water and air pollution.

Workers are poorly paid; many work for 40p an hour.

The positive and negative effects of the growth of the secondary industry in China

The government has spent $172 billion on protecting the environment.

90 per cent of the underground water in urban areas is polluted.

Many workers have left **rural** areas to work in urban areas. There is a lack of young, strong workers in rural areas.

Many urban areas suffer from pollution due to burning coal to produce energy, for example, in Linfen.

Key terms

Urban – built-up areas with housing and industry.

Rural – countryside areas with fields and woods.

Exam tip

Learn specific points about places and figures to gain extra marks in the exam.

Check your understanding
Tested

Produce a table of the effects of growth using the diagram above. Categorise the effects into social, economic and environmental.

Go online for answers
Online

Exam practice

Outline the effects on China of a growth in the secondary sector. **(3 marks)**

Answers online
Online

Exam tip

Read the question carefully. It may be asking for **reasons** for growth or **effects** of growth. Be sure you know the difference between them and give the required information.

Reasons for the growth of the tertiary sector in the UK since 1970

1 A rise in demand for services linked to disposable incomes

- The money that the average household had to spend on non-essential items doubled between 1987 and 2006.
- This has led to a rise in demand for services such as **beauticians** and health clubs. In 2000, 4 million people were members of fitness centres. By 2012, this figure had risen to 6 million.

2 The development of new technologies

- Large numbers of people are employed in telecommunication and computing providing services such as internet banking.
- Shops selling mobile phones and computers are now common on the high street.
- Many people also work at **call centres**; the number of employees rose from 350,000 in 2000 to 950,000 in 2008.

3 Decrease in employment in the primary and secondary sectors

- Fewer people are employed in the primary and secondary sectors which means proportionately more people are employed in the tertiary sector.

4 Demographic changes

- The average age for a woman to have a child in the UK has risen to 29 years old. This means that people spend more time and money socialising and have their families later in life, causing a rise in demand for service industries.
- Approximately 35 per cent of the UK population are over 50 and 20 per cent are over 65. This has led to a large 'wealthy' retired population in the country. The spending of these people is known as the '**grey pound**'. They spend money on gardening, holidays and their grandchildren.

Check your understanding

Tested

How does the growth in health club membership cause a growth in the tertiary sector?

Go online for answers

Online

Key terms

Disposable income – money left over each month from wages after all essentials have been paid for.

Beautician – someone who provides treatments such as massages or facials.

New technologies – the invention of the internet and mobile phones.

Call centres – offices where people answer the phones to deal with complaints or to take orders for goods.

Ageing population – a large number of people in the population are over the age of 50.

Demographic – population.

Grey pound – money spent by retired people.

Exam practice

Explain the factors which have caused a growth in the tertiary sector in the UK since 1970. **(3 marks)**

Answers online

Online

Economic locations

What factors affect the location of primary industry?

Revised

Location of raw material
Kaolin is only found in the south-west of England.

Market
There was a demand for kaolin by Josiah Wedgwood for the production of porcelain. This meant that there was a market for mined kaolin. By 1860, 65,000 tons were being mined each year.

Primary industry: extraction of kaolin (china clay), St. Austell, Cornwall

Transport
The kaolin was moved to Par on the south coast of Cornwall by train. Ships took the kaolin to Liverpool where it was transported by barge to Winsford in Cheshire and then by packhorse to Stoke-on-Trent.

Check your understanding
Explain why each of the factors is important for the location of primary industry.

Go online for answers

Exam practice
Which is the most important location factor for a primary industry? **(3 marks)**

Answers online

Online

What factors affect the location of secondary industry?

Revised

Tradition
There is a tradition of car manufacturing in the area. The workers are skilled and there are many suppliers of component parts.

Land
The land is a greenfield site with 280 hectares of flat land. It is cheaper land on the edge of the city of Derby.

Secondary industry: Toyota carfatory at Burnaston, near Derby

Transport
It has excellent transport routes. It is on the junction of two main trunk roads, the A50 and A38.

Area
It is located close to villages such as Findern and to the Peak District, so is attractive to managerial workers.

Exam practice
Explain the factors that affect the location of secondary industry. **(3 marks)**

Answers online

Online

Check your understanding
Explain why each of the factors is important for the location of secondary industry.

Tested

Go online for answers

Online

What factors affect the location of tertiary industry?

Revised

Demand
There is a large housing estate nearby called Salisbury village. The inhabitants form a large proportion of the club's clients. The club is close to an extensive indoor shopping area called The Galleria. People could shop and go to the gym in one trip.

Location
It is located on Hatfield Business Park. There are many large firms on the business park such as T-Mobile and Ocado. The workers from these businesses may use the health club.

Tertiary industry: David Lloyd Health Club, Hatfield

Transport /accessibility
The club is near an excellent transport system – very close to Junctions 3 and 4 of the A1(M) motorway, giving easy access to a large number of potential users. It is within walking distance of Hertfordshire University, which is a potential source of clients.

Check your understanding

Tested

Explain why each of the factors is important for the location of tertiary industry.

Go online for answers

Online

The benefits and costs of deindustrialisation in rural areas

Revised

Key term
Deindustrialisation – areas are no longer used for industrial purposes.

Deindustrialisation can have many costs and benefits for rural areas.

Example	Costs	Benefits
South Wales: Costs and benefits of the extraction of raw materials especially coal has left the landscape scarred with many waste heaps. In 1966, the Aberfan disaster occurred.	A landslide occurred from a coal waste heap after heavy rainfall and tons of material fell onto the village school killing 144 people, 116 of them children.	The land is now used for agriculture and leisure. Visitors to the area would be unaware of its industrial past.
Reading in Berkshire: The extraction of sand and gravel around Reading in Berkshire has left many quarries full of water.	The extraction left many dangerous water-filled quarries. Many jobs were lost when the quarries were closed so other industries needed to be developed to provide jobs for the previous quarry workers.	These quarries are now being used in a number of different ways. Copthorne Hotel was built next to a 10-acre lake. The hotel has many sporting facilities including water sports. It provides a number of jobs for the local community. Green Park, which is a science park, covers 70 hectares and employs 7000 people.
Eden Project, Cornwall: The extraction of china clay in Cornwall left many pits full of water. One of these pits has become the Eden Project.	The pit is 60 m deep and covers an area equivalent to 35 football pitches. When the quarries closed, a number of people lost their jobs and the local government lost income from the quarry owners.	The pit has been totally transformed into a tourist attraction with landscaped walks, a huge diversity of plants and two enormous pods. The Eden Project employs 500 staff and provides jobs for approximately 3000 other people in restaurants, hotels and suppliers of products. It is estimated that since it opened in 2001 it has contributed £1 billion to the Cornish economy.

Exam practice

Explain the benefits and costs of deindustrialisation in rural areas.

(6 marks)

Answers online

Online

Factors affecting settlements

Functions of settlements

Revised

Key term

Functions of settlement – the reason why a settlement first develops in an area.

Residential
A major function of many settlements is to give people a place to live. In some settlements this is its main function. They are often found close to larger towns or cities where the inhabitants work. Another type of residential function is one that provides for retired people. A number of settlements along the south coast of England, such as Eastbourne, have taken on this function.

Administrative
The main function of a number of settlements is as an administrative centre. These settlements are often county towns that employ a large number of people as civil servants and are centres of local government.

Market centres
A market centre's main function is to provide services for the local area. They are often found in fertile farming areas and in the past farmers would have brought their produce to sell in the town. These settlements have good transport links and many were centred on a bridging point over a river, giving access to both sides of the river. The market centre contained many services and usually a market place where the weekly market would be held.

Strategic
Strategic settlements were built in locations that used physical geography to protect them from attack:
- on top of hills for defensive purposes
- on the inside of meander bends
- beside a gap in a range of hills
- on an island in a river.

Functions of settlements

Industrial
The main function of industrial settlements in the past was to provide jobs in secondary industry. They were located on coalfields and had good access to railways and canals for transport. Many of them were found on coalfields in north Staffordshire, such as Kidsgrove near Birmingham.

Tourist resorts
Tourist resorts developed with the arrival of the railways which meant that people were able to travel around the country more easily.
- Some developed in coastal locations in settlements like Brighton on the south coast and Blackpool on the north-west coast. They developed to provide for the population who want a place to visit for recreational purposes by the sea.
- Others grew around spa towns such as Bath.
- More recently settlements in National Parks have developed as tourist resorts.
- Major cities are now tourist resorts.

Check your understanding

Tested

1 Give an example for each of the strategic settlements.
2 Why do tourists visit spa towns and major cities?

Go online for answers

Online

Exam practice

Outline three functions of settlements. **(6 marks)**

Answers online

Online

How the functions of a UK settlement have changed over time

The functions of a settlement can change over time as society develops. Aberfan is a small settlement in South Wales which has seen its function change over time.

An **agricultural settlement** grew in the bottom of the valley close to the River Taff, approximately 5 miles from Merthyr Tydfil.

↓

Merthyr Vale coal mine was opened in 1875. This changed the main function of the village and the village became an **industrial settlement**. There were still farms and agricultural workers in the village but the majority of the population worked in the mine.

↓

In 1989 the coal mine was closed and the function of the settlement changed to being a **residential settlement**. It is a commuter village with the people who live there working in local towns and cities such as Merthyr Tydfil. It is also a retirement settlement as the age structure of the population shows a larger number of people in the retirement age groups.
There are still a number of farms close to the village but the main function is residential.

Check your understanding
Tested

List the functions that Aberfan has provided over the years.

Go online for answers
Online

Exam tip

When the command word **compare** is used, you should state the similarities between the photographs or figures that have been given. However, examiners at GCSE will also credit comments about the differences (contrasts).

Exam practice

For a named example, describe how the functions of a settlement have changed over time.

(4 marks)

Answers online

Online

Changes to rural communities

Counter-urbanisation

Changes to rural areas can be positive and negative.

Environmental changes

- Many of the migrants still work in urban areas, causing pollution.
- Villages become ghost towns during the day.
- Old derelict farm buildings are turned into habitable dwellings which adds to the aesthetic value and community well-being.

Key term

Counter-urbanisation – the movement out of cities to rural areas or smaller urban settlements. This process has been happening in HICs for the last 50 years.

Social changes

- The traditions of the village are not valued by the newcomers.
- Many church parishes have been amalgamated as the 'newcomers' do not go to church.
- Local schools have an increase in pupils and are able to stay open.

Economic changes

- House prices in rural areas may rise as demand increases. This may mean that local people cannot afford to buy a house and have to move away from their local area.
- Many of the migrants do not support local businesses and do their shopping in the urban areas where they work.
- Some local services are supported such as public houses, local tradesmen (for example, builders).

Demographic changes

The people who tend to move to rural areas are the more affluent. They either have a young family or are retired.

Counter-urbanisation example: Austrey in Warwickshire

- People moved out of the city of Birmingham and the local town of Tamworth into this village during the 1970s. It caused a growth in population from 300 in 1961 to 1000 in 2001.
- The environment of the village changed with a number of new housing estates being built such as St Nicholas Close and Elms Drive on previous farmsteads. There used to be eighteen farms in the village; only two are left. The buildings of the others have been converted into houses meaning that the village has lost some of its original character.
- The village school has opened on a new site with 120 children on roll; in 1961 there were sixteen children.
- The village pub, 'The Bird in Hand' is thriving and has become a meeting place for the local community.
- Many of the people who live on the new estates work in Birmingham and most families have two cars, although the newcomers do use the village shop for their provisions.

Check your understanding
Tested

Categorise the changes that occurred to Austrey under the following headings: demographic, economic, social and environmental changes.

Go online for answers
Online

Exam practice

Explain the changes that have occurred to rural communities due to counter-urbanisation. Use an example in your answer.   **(4 marks)**

Answers online
Online

Depopulation of remote rural areas

Demographic changes

There has been a decline in population for the age bands up to 40 in all remote rural areas. Young adults leave the area which means that there are fewer young children. The population then develops an older structure.

Economic changes

As the population becomes older there will be less money going into the running of public services such as waste treatment and water. This can cause problems for local councils.

The economy of the area also decreases as less money is going into the economy because fewer people of working age live there.

Environmental changes

In some remote rural areas there are signs of neglect and derelict buildings which can be unappealing. The decrease in population is, however, a bonus for the wildlife of the area.

Social changes

This movement out of remote rural areas has meant that the population who remain there have seen a decline in service provision. In 2001 there were 600,000 people living in what can be classed as remote rural areas; 45 per cent of these people did not live within 4 km of a doctor's surgery or a post office and were also without a bus service.

This situation will continue to worsen as many post offices are closing in rural areas in the UK as they are not profitable. In Cornwall 25 per cent and in Devon 22 per cent of post offices are set to close, while the county average for the UK is 18 per cent.

The decline in rural services has also seen the closure of many primary schools such as Satterthwaite and Rusland School and Lowick School, both near Ulverston in the Lake District. They were closed in 2006.

> **Key term**
>
> **Depopulation of rural areas** – people moving away from the countryside.

> **Exam tip**
>
> Don't be caught out. Learn points for social, economic, demographic and environmental changes.

> **Exam tip**
>
> **Foundation Tier**
>
> For questions that ask for examples, your answer will be marked as follows:
>
> - Each point will receive a mark.
> - If your answer does not contain a specific point about an example you will lose 1 mark.
>
> **Higher Tier**
>
> - If the command word is **outline** or **describe**, these questions will usually be marked out of 4 marks.
> - Each point will receive a mark.
> - If your answer does not include specific points about an example you will only receive 2 marks.
> - If examples are asked for and you only give one you will lose 1 mark.

> **Exam practice**
>
> What is meant by the term rural depopulation? **(2 marks)**
>
> **Answers online**
>
> Online

Changing land use in urban areas

Land use in urban areas ──────────────────── Revised

Land use in urban areas in the UK has shown a dramatic change over the last 30 years. This is due to two significant trends:

- an increased demand for housing by the UK population
- deindustrialisation – manufacturing has moved from urban areas in the UK to LICs where production costs are much lower.

Social

- People are now marrying later in life – the average age has gone up from 24 in 1960 to 30 in 2010.
- People are having fewer children, later in life, which has also impacted on the type of houses that are demanded. More flats and smaller houses are now being built.
- There has also been a rise in the number of divorces which means that a family are not living as a group but are living in two different dwellings.
- Many people now live on their own or with their spouse until they are in their 70s and 80s – this means that more houses are needed for the younger generation.

Economic

- The population is wealthier, therefore young people can afford to rent flats at an earlier age and no longer have to live with their parents.
- Until recently people were encouraged to buy their own properties because 100 per cent mortgages were available.

Political

- The population of the UK is increasing. It is predicted to rise by 4.1 million between 2001 and 2021.
- The government has promised that 3 million new homes will be built by 2020. The growth will take place in certain areas of the country. One of the developments in the south-east is in Bracknell.
- There has been a large influx of EU nationals since the relaxation of borders between EU countries.

Check your understanding ──────────────────── Tested

List the social reasons for the increase in demand for housing.

Go online for answers ──────────────────── Online

Exam practice

Using examples, explain changes in land use that have occurred in urban areas in the UK. **(4 marks)**

Answers online

Online

Exam tip

You could be asked questions that require recall of knowledge. You should learn at least one specific social, economic and political reason.

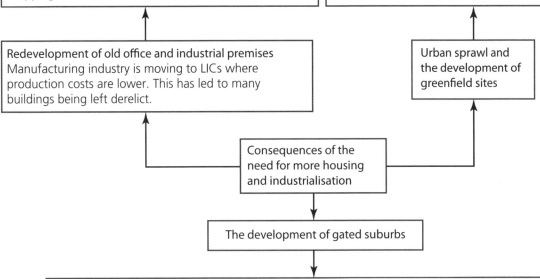

A large site in Norwich comprising 17 hectares is located south-east of Norwich city centre on the banks of the River Wensum, close to the railway station and the football ground at Carrow Road. The site has been **redeveloped** over a number of years into an entertainment complex including a fourteen-screen cinema, a large shopping centre and over 200 residential units.

Bracknell in Berkshire is seeing significant growth with a large new housing development on a **greenfield site** to the west of the town. Peacock Farm estate is situated next to the A329M and is close to the M4 for easy access for residents. The development includes 14,000 new homes, 91 acres of country parkland, a doctor's surgery and two primary schools.

Redevelopment of old office and industrial premises
Manufacturing industry is moving to LICs where production costs are lower. This has led to many buildings being left derelict.

Urban sprawl and the development of greenfield sites

Consequences of the need for more housing and industrialisation

The development of gated suburbs

Some areas have seen the development of 'gated suburbs'. This is the concept which originated in countries such as South Africa to protect residents. A number of housing developments, for example in the south-east and the north-west, now have gates to protect them from perceived threats such as burglars.

Check your understanding

Tested ☐

Outline what is meant by a gated suburb.

Go online for answers

Online ☐

Key terms

Redevelopment – when buildings in a city, which are no longer of use, are demolished and replaced with buildings that are in current demand.

Renewal – when old buildings are renovated and brought up to date, combining the best of the old with the new.

Brownfield site – an area within a city, which is no longer used. It may contain old factories and housing, or it may have been cleared ready for redevelopment.

Greenfield site – an area on the edge of the city, which has never been developed in any way.

Exam tip

You may be asked questions about the consequences of the increase in demand for housing with reference to a photograph or other resource. In this case you will have to apply your knowledge about the consequences to the area shown.

Exam practice

Outline the difference between a brownfield and a greenfield site. **(3 marks)**

Answers online

Online ☐

The advantages and disadvantages of brownfield sites

Revised

Advantages	Disadvantages
Planning permission is easier to get; the government is actively encouraging the use of these sites.	Complete environmental survey needed because of past usage is costly and time consuming.
Infrastructure, such as gas, electricity and water, is already present.	Brownfield sites have to be cleared and in some cases decontaminated, which adds to the construction costs.
Sites are easier to market because of access to entertainment and other facilities.	Cities may have social problems, such as anti-social behaviour and crime, as well as higher levels of pollution and congestion which could make marketing more difficult.
No building on greenfield sites so lessens urban sprawl.	Land costs are higher as it is closer to the city centre.

The advantages and disadvantages of greenfield sites

Revised

Advantages	Disadvantages
Originally unoccupied therefore developers can build as they wish.	Infrastructure, such as gas, electricity and water, will not be present.
Plenty of space for car parking and landscaping to improve the working environment.	Urban sprawl uses up green spaces on the edge of urban areas.
Cheaper land due to it being further from the city centre.	It is more difficult to get planning permission as the government tends to be against it.
Lower construction costs as there is nothing to knock down or renew.	Building could disturb natural habitats and wildlife.
Easy to market to potential buyers because of pleasant environment.	Living on the edge of the city may increase the commute for some people.
Access to the development is easier as roads are not congested.	People may not want to live away from the city centre because of their social life.

Check your understanding

Tested

1 Explain the difference between redevelopment and renewal.
2 Outline three advantages and three disadvantages of developing on brownfield sites.

Go online for answers

Online

Exam practice

State the advantages of building on a greenfield site.   (3 marks)

Answers online

Online

Attitudes of central and local government to development

Housing minister	'Due to the increase in the population we have put policies in place to cope with the increased demand for housing. There will be growth in certain areas of the south-east such as Ashford in Kent.'
Bracknell Forest Councillor	'We were very sorry to lose the greenbelt land at Peacocks Farm but new development is good for the town, bringing in more people and businesses.'
Shadow housing minister	'The government must realise that there has been a change in the population structure with people living longer and marrying later. This means that more smaller housing needs to be built.'
Green Party member of Bracknell Forest Council	'This continued development on the greenbelt will mean that we will soon be joined up with Wokingham. There are other places that could be developed within the town which would protect the land on the edge.'

Check your understanding
Tested

Which are attitudes of central government and which are attitudes of local government?

Go online for answers
Online

Attitudes of individuals and organisations to development

Revised

Bracknell resident	'When the Met Office closed a lot of jobs were lost but the building was old and 'run down'. The new apartments are stylish and have made that area of the town look more up to date.'
Peacocks Farm resident	'The new estate is very convenient for the M4. I love living on the outskirts of the town – it gives me the best of both worlds.'
Norwich resident	'The renewal and redevelopment that has taken place near Carrow Road has really improved that part of Norwich. There are now entertainment facilities and the riverside is great in the summer for a walk along the new paths.'
Factory owner, Norwich	'It was a shame to see the factory go as it had been in the family for years but we were not making money. The renewed buildings look great and blend in well with the area.'

Exam tip
When you are learning the attitudes to development, remember to learn who made the comment.

Rapid growth in LICs

Reasons for the rapid growth of urban areas in LICs

Urban areas in LICs have experienced a rapid growth since the 1950s.

There are two main reasons for this rapid growth:
● the migration from rural to urban areas
● a high natural increase in population in urban areas.

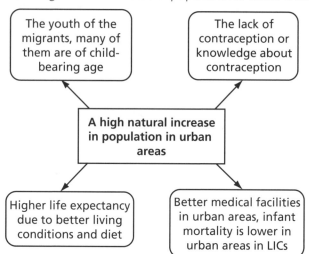

Lack of jobs in rural areas because of population growth and mechanisation		Salaries are lower in rural areas

The migration from rural to urban areas

The perception of a better life, including education		The development of TNCs and other industry providing jobs in urban areas

The youth of the migrants, many of them are of child-bearing age		The lack of contraception or knowledge about contraception

A high natural increase in population in urban areas

Higher life expectancy due to better living conditions and diet		Better medical facilities in urban areas, infant mortality is lower in urban areas in LICs

The effects of rapid growth on a LIC urban area – Cairo

Air pollution

- Air pollution from the 2 million cars and the 200,000 motorbikes.
- In the industrial quarter Shoubra al-Kheima, where many people live close to their work, 37 per cent of the residents suffer from lung problems.
- The Sun's rays are blocked by smog on the most polluted days which means that many children suffer from a deficiency of vitamin D.

Water pollution and problems

- 23 per cent of the population of Cairo do not have access to a fresh water supply.
- 25 per cent of the population are not connected to the public sewage system.

Land pollution

- The inhabitants of Cairo produce 10,000 tonnes of solid waste a day. Only 60 per cent is collected; the rest is left to rot in the streets.
- Large toxic stockpiles of hazardous waste, as much as 50,000 tonnes, from industry which has accumulated in Helwan, Shoubra and Embaba.

Noise pollution

- Noise pollution from the 2 million cars and the 200,000 motorbikes.
- Loud speakers calling Muslims to prayer.
- Noise of nightclubs on the River Nile. It is particularly bad in the Saraya Al Gezira district.

Housing problems

- Approximately 60 per cent of Cairo's population live in shanty type dwellings. The most famous of these is the 'City of the Dead', or Arafa (cemetery) as it is called by the local residents. This four-mile-long cemetery in eastern Cairo is where people live and work among their dead ancestors.
- The government has responded to the housing problem by building cities on the edge of Cairo in the desert. Two of these are 6th of October and 10th of Ramadan. Many Cairo residents, however, want to stay in the city where their jobs are.

Chapter 14 Population Change

Population growth and distribution

Global population change
Revised

The world's population has grown very quickly over the last 200 years from 1 billion in 1800 to the present population which is over 7 billion. This growth in population is not even, with the majority of the growth being in Asia and Africa. The death rate in countries in these continents has decreased dramatically while the birth rate has remained high causing a rapid growth in population.

Check your understanding
Tested
Describe the changes shown on the graph.

Go online for answers
Online

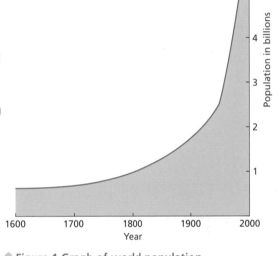

↑ Figure 1 Graph of world population

Global population distribution
Revised

Reasons for the distribution of global population

Factor	Sparsely populated areas	Densely populated areas
Climate	Extremely hot or cold climates such as the Sahara Desert or Siberia.	Temperate climates where the population can grow crops such as the UK.
Soil	Infertile soils where crops cannot be grown such as mountainous areas, for example, the Himalayas.	Fertile soils where crops are grown such as river valleys, for example, the Ganges Valley.
History of settlement – resources	Few resources for industry, for example, the Sahel.	Many resources for industry such as coal, for example, France.

Key terms
Population distribution – how people are spread across an area.
Sparsely populated – few people live in an area.
Densely populated – many people live in an area.

Exam tip
You may be asked to describe a distribution on a map.
- You should start with **general** points.
- Your answer should then become more **specific**.
- If data is not requested you should still include some as you may be given credit.

Exam practice
Explain the reasons for the distribution of global population. **(4 marks)**

Answers online

Online

Reasons for changes to birth and death rates

Revised

Medical
- New treatments to beat cancer are being invented. People are also becoming more aware of the link between unhealthy living and early death. Both of these mean that the death rate is falling in HICs.
- Better access to health care including inoculations for childhood diseases has decreased death rate so fewer children are being born.

Social
- People now live in healthy conditions with sewage systems and tap water. This will lower death rates.
- If women receive an education they become aware of ways to control fertility. It also increases their choices in life. This leads to a lower birth rate.
- The average age of marriage in the UK has risen from 24 in 1960 to 30 in 2010. Birth rates will be lower as there is less time to have children.

Reasons for changes to and death rates

Economic
- In the UK, children cost a lot of money to feed and clothe. Therefore fewer children are being born because couples prefer to provide more for one or two children.
- Poverty affects the death rate in parts of the UK. Areas which have high levels of poverty also have high death rates.

Political
- China has used policies to decrease its birth rate, for example the 'one-child' policy, whereas Singapore has used incentives to increase its birth rate.
- Governments in LICs have invested large sums of money to provide the population with clean drinking water to lower the death rate.

Check your understanding

Tested

List two reasons for changes in birth rate and two reasons for changes in death rate.

Go online for answers

Online

Exam practice

Outline the reasons for the changes to birth rate. **(4 marks)**

Answers online

Online

Key terms

Birth rate – the number of people born per 1000 of the population in a year.

Death rate – the number of people who die per 1000 of the population in a year.

The characteristics of the demographic transition model

Revised

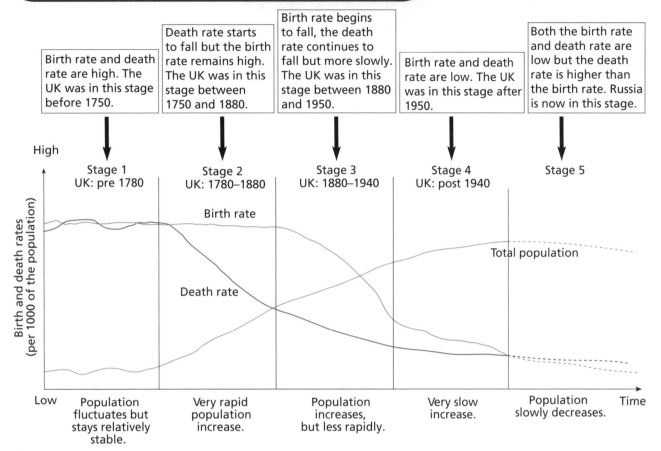

Birth rate and death rate are high. The UK was in this stage before 1750.

Death rate starts to fall but the birth rate remains high. The UK was in this stage between 1750 and 1880.

Birth rate begins to fall, the death rate continues to fall but more slowly. The UK was in this stage between 1880 and 1950.

Birth rate and death rate are low. The UK was in this stage after 1950.

Both the birth rate and death rate are low but the death rate is higher than the birth rate. Russia is now in this stage.

High

Stage 1
UK: pre 1780

Stage 2
UK: 1780–1880

Stage 3
UK: 1880–1940

Stage 4
UK: post 1940

Stage 5

Birth and death rates (per 1000 of the population)

Birth rate

Death rate

Total population

Low

Population fluctuates but stays relatively stable.

Very rapid population increase.

Population increases, but less rapidly.

Very slow increase.

Population slowly decreases.

Time

↑ Figure 2 The demographic transition model

Check your understanding
Tested

Describe the changes to the birth rate shown on the demographic transition model.

Go online for answers
Online

Exam practice

Explain the changes to the death rate in Stage 2 of the demographic transition model. **(4 marks)**

Answers online

Online

Exam tip
Foundation Tier

For questions that ask for examples, each point will receive a mark. If you do not include an example, you will lose 1 mark.

Higher Tier

- If the command word is **outline** or **describe**, these questions will usually be marked out of 4 marks.
- Each point will receive a mark. If your answer does not include specific points about an example you will only receive 2 marks.

Exam tip

You could be asked about government attitudes towards population change. Revise the changes for one country. For example, Singapore in the 1970s encouraged professional couples to have children by giving them monetary incentives. The policy has now changed and all of the population can get the incentives.

The physical and human factors affecting the distribution and density of population in China and the UK

Revised

China

The west of the country is sparsely populated. In this area are the Atai Mountains where it is too wet and cold to grow crops.

The east of the country is densely populated. People live in coastal areas due to good trade links.

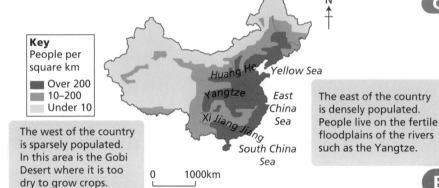

Key
People per square km
■ Over 200
■ 10–200
□ Under 10

Huang He — Yellow Sea
Yangtze — East China Sea
Xi Jiang Jiang — South China Sea

0   1000km

The west of the country is sparsely populated. In this area is the Gobi Desert where it is too dry to grow crops.

The east of the country is densely populated. People live on the fertile floodplains of the rivers such as the Yangtze.

⬆ Figure 3 How physical and human conditions have influenced the density and distribution of population in China

The UK

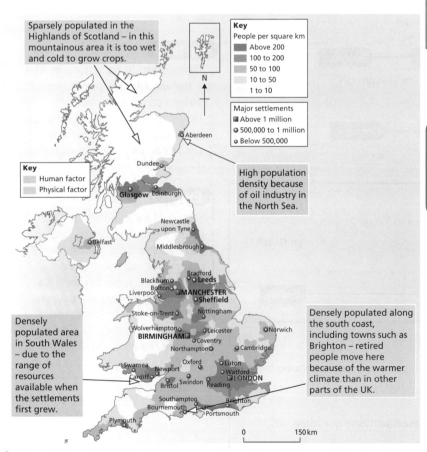

Sparsely populated in the Highlands of Scotland – in this mountainous area it is too wet and cold to grow crops.

Key
People per square km
■ Above 200
■ 100 to 200
■ 50 to 100
□ 10 to 50
□ 1 to 10

Major settlements
■ Above 1 million
◉ 500,000 to 1 million
● Below 500,000

Key
■ Human factor
□ Physical factor

High population density because of oil industry in the North Sea.

Densely populated area in South Wales – due to the range of resources available when the settlements first grew.

Densely populated along the south coast, including towns such as Brighton – retired people move here because of the warmer climate than in other parts of the UK.

Aberdeen, Dundee, Glasgow, Edinburgh, Newcastle upon Tyne, Belfast, Middlesbrough, Blackburn, Bolton, Bradford, Leeds, Liverpool, MANCHESTER, Sheffield, Stoke-on-Trent, Nottingham, Wolverhampton, Leicester, BIRMINGHAM, Coventry, Norwich, Northampton, Cambridge, Swansea, Oxford, Luton, Cardiff, Newport, Watford, LONDON, Bristol, Swindon, Reading, Southampton, Brighton, Bournemouth, Portsmouth, Plymouth

0   150 km

⬆ Figure 4 How physical and human conditions have influenced the density and distribution of population in the UK

Check your understanding

Describe the factors that have influenced population distribution in China.

Go online for answers

Exam practice

Explain the physical factors which have affected the distribution of population in China.   **(4 marks)**

Answers online

Online

Exam tip

If the question asks generally for factors affecting population distribution, remember to write about human and physical factors.

Two countries with contrasting population problems

China

Why did China need to reduce its birth rate?

In 1979 a quarter of the world's population lived in China, with 66 per cent of the population being under the age of 30. This meant that in the future the population would increase even more when these people had children.

The government brought in a population policy which they hoped would promote economic growth and improve the living conditions of the majority of the population. The policy is known as the one-child policy – couples were allowed to only have one child. They were given a 'one-child certificate' which entitled them to a number of benefits which would improve their standard of living.

What has been done to decrease China's birth rate?

The one-child policy has a number of incentives and disincentives:

Incentives of the policy	Disincentives of the policy
Cash bonuses	Couples who work for the government were sacked from their jobs if they had more than one child.
Free education	If they had two children they lost all of their privileges.
Free medical care	People were monitored by 'granny police'.
Preferential treatment over housing	Couples had to ask permission to have a child.

Check your understanding   Tested

State why China has introduced policies to decrease its birth rate.

Go online for answers   Online

Singapore

Why did Singapore need to increase its birth rate?

In the late 1980s, the government of Singapore realised that the country's low birth rate would mean that they would not have enough workers. In 1987 they introduced the 'three or more policy' to encourage people to have more children.

What has been done to increase Singapore's birth rate?

The 'three or more policy' has a number of incentives and disincentives as shown on the right.

Incentives of the policy	Disincentives of the policy
Parents receive a gift of $3000 from the government for each of their first two children and $6000 for each of their next two children.	Couples with one or no children are only allowed to buy a three-room flat.
Fathers receive three days of paternity leave on the birth of the first four children.	They are not allowed to choose the school for their children so their education might suffer.
Mothers are allowed three months of maternity leave.	They do not receive financial packages offered by the government.

Check your understanding   Tested

State one incentive and one disincentive that Singapore has used to try to increase its birth rate.

Go online for answers   Online

Exam practice

Give two reasons why China wanted to decrease its birth rate. (4 marks)

Answers online   Online

Exam tip

You need to learn at least three incentives and disincentives for each country. Try to learn specific points (facts and figures) as well as general points.

Characteristics of population

The characteristics of population on a local scale including age, gender, ethnic, religious and occupational structure

Revised

Key terms

Census – a count of the number of people in a country. It provides statistics about a number of population characteristics for all the households in a country. The last census took place in the UK in 2011.

Age – the age structure of the people in an area.

Gender – the number of males and females in the area.

Ethnic group – a group of people who have a common national or cultural tradition.

Religious structure – the religion (faith) of the people who live in the area.

Occupational structure – the types of jobs that are done by people in the area.

The **census** is a record of all the people living in a country. It contains information on many different characteristics including **age**, **gender**, **ethnic group**, **religious** and **occupational structure**. The local council will look at the results of the census in their area to help them to provide services for the community.

You could be asked to describe, understand or interpret census data in the same way that your local council does. In Figure 5, the council of Area A would need to increase its provision in education because of the large numbers of young people.

Age structure

Age	Area A: % of population	Area B: % of population
0–5	25	5
6–15	15	10
16–29	15	15
30–44	25	20
45–64	15	25
65 and over	5	25

↑ **Figure 5** Age structure of two different areas

Exam tip

You need to learn all the terms relating to census data and practise interpreting census data.

Check your understanding

Tested

Interpret the age structure census data for Area B in Figure 5.

Go online for answers

Online

Comparison of population pyramids for three countries at different levels of development

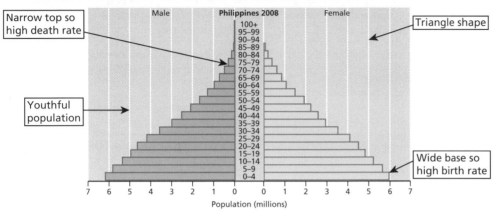

Narrow top so high death rate

Youthful population

Male · Philippines 2008 · Female

Triangle shape

Wide base so high birth rate

Population (millions)

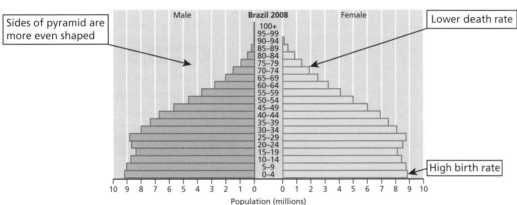

Sides of pyramid are more even shaped

Male · Brazil 2008 · Female

Lower death rate

High birth rate

Population (millions)

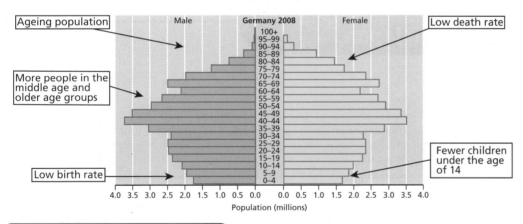

Ageing population

More people in the middle age and older age groups

Low birth rate

Male · Germany 2008 · Female

Low death rate

Fewer children under the age of 14

Population (millions)

Check your understanding

Tested

Compare the characteristics of the three pyramids, looking at shape, birth rate, death rate, growth rate, stage of demographic transition model, age structure and the future.

Go online for answers

Online

Exam practice

Compare the shapes of the pyramids for Germany and the Philippines. **(4 marks)**

Answers online

Online

Exam tip

On the Higher Tier paper you may be asked to relate a population pyramid to the demographic transition model.

The consequences of a youthful and an ageing population

Revised

Consequences	Youthful population	Ageing population
Health services	There will be greater demand for midwives and hospital care for babies and mothers.	Greater need for care homes and specialist nursing for the elderly.
Education services	There will be a great demand for education services.	There will be less money for education due to the demand for elderly services such as care homes.
Housing provision		More houses will be needed because people are living longer.
Pensions provision	Less demand for pensions because of younger population.	The pension age will have to increase from 65 to 68 in 2046.
Other	There is a large workforce. There are many children to look after their parents in the future so less care needed for the elderly in the future.	The elderly work without pay in charity shops. The retired go on more holidays. So there will be more jobs in the leisure industry.

Exam practice

1 Explain the consequences of a youthful population. **(4 marks)**
2 Explain the consequences of an ageing population for Japan. **(6 marks)**

Answers online

Online

Exam tip

You might be asked for just negative or positive consequences so make sure that you learn both.

Check your understanding

What is meant by the greying yen?

Go online for answers

The advantages and disadvantages of an ageing population in Japan

Revised

Health care
- There are more people living in nursing homes.
- A new health insurance scheme for the over-75s was introduced in 2008. It was nicknamed the 'hurry up and die' scheme. After a patient has been in hospital for 100 days, the fee the hospital receives from the government goes down. It is hoped that this will shorten stays in hospital.

Workforce
- The number of people of working age is decreasing.
- The staff of Tokyo's subway are mainly pensioners.
- In 1990 there were almost six people of working age for each pensioner. By 2025 that number will be down to two.

Advantages and disadvantages of an ageing population in Japan

Pensions
- The age for retirement in Japan is rising from 60 to 65 by 2030.

Technology
- Many pensioners live on their own and have children who worry about them. This has led to a number of gadgets being invented. They include an online kettle which sends e-mails to up to three people when it is switched on.

The greying yen
- Retired people in Japan are spending money on luxury goods such as holidays.
- This will lead to greater employment and a growth in the economy.

Chapter 15 A Moving World

Population movement

Section B – Remember that you have only been taught one of these topics. If you have been taught A Tourist's World, turn to page 113.

Different types of population movement
Revised

Key terms

Migration – the movement of people from one place to another.

Immigration – movement into a country.

Emigration – movement out of a country.

Short-term population movements – movements of people for a short period of time such as for holidays, commuters and university students.

Exam tip

When the command word **compare** is used, you should state the similarities between the photographs or figures that have been given. However, examiners at GCSE will also credit comments about the differences (contrasts).

Check your understanding
Tested

What is the difference between immigration and emigration?

Go online for answers
Online

Classification of migration	Definition
National OR	Movement from one area of a country to another
International	Movement between countries
Long-term OR	Usually a permanent movement, such as for retirement
Short-term	For a short period of time, such as going to university in a different city
Voluntary OR	People move because they want to, perhaps to improve their lives
Forced	People move because they have to
Legal OR	The government of a country knows about the **migration**
Illegal	The government of the country does not know about the migration

Exam practice

Explain the difference between the following terms:

1 voluntary and forced migration
2 national and international migration (4 marks)

Answers online
Online

Check your understanding
Tested

Try to think of an example for each of the classifications in the table above.

Go online for answers
Online

Flows of population

The main migration flows into and within Europe since 1945

Key terms

Iron Curtain – countries in Eastern Europe which did not allow movement to the west of Europe.

Colonies – countries that have been conquered and are ruled from another country.

Commonwealth countries – countries that were conquered and ruled by the UK.

European Union (EU) – a group of countries in Europe which have no restrictions on travelling and work between them.

Flows into Europe	Flows within Europe
Germany after the war needed people to rebuild the country. Many workers migrated from Turkey to Germany.	After the fall of the **Iron Curtain** in 1989 there was a large movement of people from east Europe to west Europe.
A number of European countries received migrants from their **colonies**. • France received migrants from countries in North Africa such as Algeria and Senegal. • The UK received migrants from India and the Caribbean.	The growth of the **EU** has led to an increase in migration between member states, in particular new member countries in east Europe moving to Germany, France and the UK. In 2008, 700,000 Poles were working in the UK.
Between 1960 and 1990, a number of European countries received political migrants who were looking for safety. Spain received migrants from Argentina and Uruguay. Portugal received migrants from Brazil.	People move from northern Europe, for example the UK, to countries in southern Europe such as Spain when they retire.

Exam tip

You may be asked to describe migration flows into or within Europe shown on a map.

- You should start with **general** points about different parts of the map, for example using compass points to locate information shown.
- Your answer should then become more **specific** using the names of countries.
- If evidence is asked for, you will lose a mark if you do not include it.
- If evidence is not requested, you should still include some as you will be given credit.

Exam practice

Describe one population flow into Europe and one population flow within Europe. **(4 marks)**

Answers online

The social and economic impacts of population movements on the host country and the country of origin

Revised

Social impacts

Polish shops on many British high streets gives variety to the culture.

Migrants are generally in their 20s and 30s. This is making the UK workforce younger.

The Cambridgeshire police force has to deal with over 100 different languages. It is difficult and expensive to find translators.

The positive and negative impacts on the host country (UK)

Economic impacts

The average migrant worker earns £20,000 per year in the UK, adding to the economy.

Some 27,000 child benefit applications have been approved. This is a burden on the welfare state.

Migrant workers do low–paid jobs such as crop picking.

Social impacts

Wages have increased in Poland which means people have a higher standard of living.

Rural areas are losing population in the 20–30 age group as they are the ones who migrate. This means that schools are closing.

The birth rate in Poland has decreased because of the loss of people in the reproductive age groups.

The positive and negative impacts on the country of origin (Poland)

Economic impacts

Scientists and researchers have each received a one-off payment of £5000 to try to encourage them to stay in Poland.

In 2007 there were many vacancies in the construction industry because of a shortage of workers.

In 2007 salaries in Poland increased by 9%.

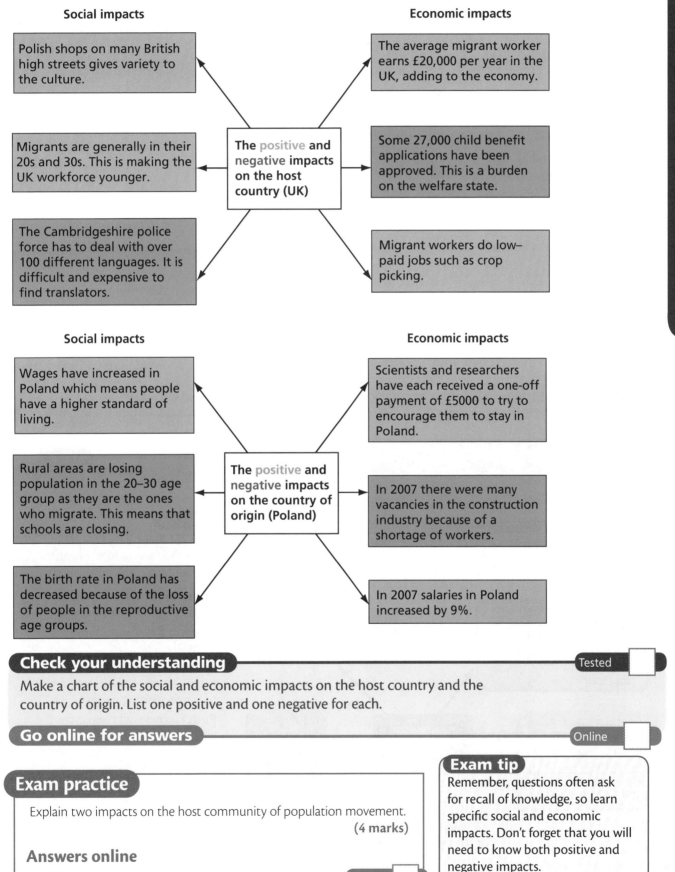

Check your understanding

Tested

Make a chart of the social and economic impacts on the host country and the country of origin. List one positive and one negative for each.

Go online for answers

Online

Exam practice

Explain two impacts on the host community of population movement.
(4 marks)

Answers online

Online

Exam tip

Remember, questions often ask for recall of knowledge, so learn specific social and economic impacts. Don't forget that you will need to know both positive and negative impacts.

Factors influencing rates of population movement

Technology

People can use the internet to:

- look for work and find accommodation in other countries
- keep in close contact with family and friends when they move away
- book flights and other forms of transportation to move easily around the world.

Transport

- Faster modes of transport have allowed people to move more quickly. People now commute from Birmingham to London by train.
- Budget airlines like Ryanair and Easyjet have flights to many small airports in Europe allowing people to move around much more cheaply.
- Major improvements have been made in road and rail services. The opening of the Channel Tunnel has made it easy to drive or use the train to travel to France. High-speed rail and motorways link the major cities in Europe.

Government policies

- People are free to move between the member countries of the EU. They just need a passport or identity card.
- The UK introduced an entry points system in 2008. Highly skilled economic migrants are still welcome in the UK as long as they reach the pass mark of 75 points.

Check your understanding

Tested

State three reasons why people move. Give one from each category: technology, transport and government policies.

Go online for answers

Online

Exam practice

Outline the factors affecting population movement. **(4 marks)**

Answers online

Online

Reasons for short-term population flow

Medical

These are short-term population flows for health reasons. This could be for medical treatments in France or for dental treatments in Hungary.

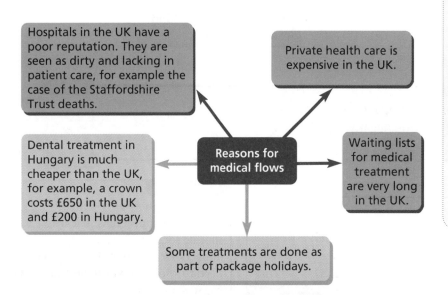

Hospitals in the UK have a poor reputation. They are seen as dirty and lacking in patient care, for example the case of the Staffordshire Trust deaths.

Private health care is expensive in the UK.

Dental treatment in Hungary is much cheaper than the UK, for example, a crown costs £650 in the UK and £200 in Hungary.

Reasons for medical flows

Waiting lists for medical treatment are very long in the UK.

Some treatments are done as part of package holidays.

Key terms

Medical flows – movement in Europe to receive better health and dental treatment.

Economic flows – moving to another area for work reasons.

Tourism flows – movement to go on holiday.

Sport flows – movement of people to either participate in or to watch sporting events.

Push factors – all the reasons why people move away from an area.

Pull factors – all the reasons why people move into an area.

Economic

These are short-term population flows for reasons related to improving standard of living. The main reason for economic migration is for a better paid job and an increase in wealth. Other migrants are seasonal workers who come to pick fruit and vegetables. Many migrants come to Lincolnshire every year to help on the fruit farms during the season.

Tourism

People move for a short period of time to have a holiday.

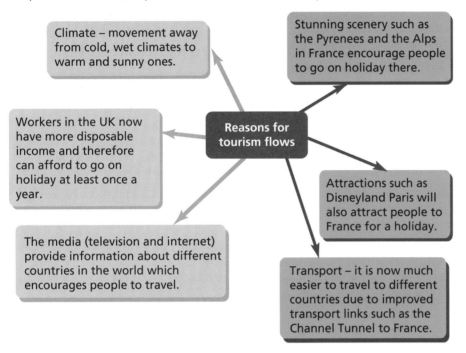

Climate – movement away from cold, wet climates to warm and sunny ones.

Stunning scenery such as the Pyrenees and the Alps in France encourage people to go on holiday there.

Workers in the UK now have more disposable income and therefore can afford to go on holiday at least once a year.

Reasons for tourism flows

Attractions such as Disneyland Paris will also attract people to France for a holiday.

The media (television and internet) provide information about different countries in the world which encourages people to travel.

Transport – it is now much easier to travel to different countries due to improved transport links such as the Channel Tunnel to France.

Sport

People move around the world to either participate in sport or to be a spectator at a sporting event.

- People want to attend sporting events, for example, thousands of people came to the UK in 2012 to watch the Olympics. Thousands of Manchester United supporters will go to Spain when their team plays Real Madrid.

- People move around the world to compete in sporting events because they want to compete against the best players in the world and to win prize money. For example, Andy Murray travels around the world competing in tennis tournaments.

Exam tip

You should know all four types of short-term movement and examples of each.

Check your understanding ———— Tested ☐

For medical and tourism flows, try to distinguish between push and pull factors.

Go online for answers ———— Online ☐

Exam practice

Explain, using examples, the reasons why people move. Refer only to medical and tourism reasons in your answer.   **(4 marks)**

Answers online

Online ☐

Retirement migration

Case study: Retirement migration to North Norfolk

Reasons for the migration	Detail and explanation
Property is cheaper	The average house price in North Norfolk is approximately £150,000 less than in London.
Stunning scenery	North Norfolk is an Area of Outstanding Natural Beauty and the coastline has been designated a Heritage coast.
Climate	North Norfolk is one of the driest areas in England with an annual rainfall of 625 mm. It also has the highest summer temperatures.
Lifestyle	North Norfolk has a slow pace of life and a lower crime rate than other areas of the UK.

Consequences of the migration	Detail and explanation
Housing	The demand for houses has increased due to the number of migrants coming into the area. This has made houses more expensive. Local people, especially young married couples, cannot afford to live in the area where they were brought up.
Provision of services	Some village services such as the local shops have benefited as there are more people in the village who will shop daily and go to the local pub. However, other services such as schools have suffered because the migrants do not have children. Some schools may have to be closed.
Population structure	North Norfolk has an ageing population. This is a burden on the council which will have to provide more medical facilities. A new hospital has been planned for North Norfolk combining a GP surgery, 24 beds and community health services under one roof. There will also be an increase in demand for care services and meals on wheels.
Village character	Some villages such as Burnham Market have lost their traditional feel as trendy shops such as Gunn Hill Clothing Company have moved into vacant premises to cater for the wealthy new migrants. Village halls now have bingo sessions and tea dances rather than youth clubs.

Case study: Retirement migration to Spain

Reasons for the migration	Detail and explanation
Climate	The temperature throughout the year in Spain is at least 10°C warmer than the UK.
Communication networks	Many budget airlines fly to Spain for approximately £50. It takes less time to fly to Spain than it does to drive to Manchester. Therefore, people can come home regularly to see their families.
Lifestyle	The lifestyle in Spain is very relaxed with a low crime rate.
Property	Property in Spain on average is cheaper than in the UK.

Consequences of the migration	Detail and explanation
Housing	There has been a lot of development along the coast of Spain. This has caused damage to the environment. In 1998 the government passed the Coastal Law which states that any property built within 106 m of the shore can be demolished.
Water	Many of the migrants move to Valencia which is a very dry area of Spain. They buy properties with swimming pools which use a lot of water. This is causing friction between the host community and the migrants.
Leisure facilities	There are many leisure facilities provided for the migrants such as tea dances. Golf courses are also being built for migrants to use – in Murcia 54 have been built in the last ten years. The demand for water is now twice the supply.
Healthcare	Many British retire to the Costa Blanca area of Spain. It is estimated that the British migrants are costing the local government £800 million a year in healthcare costs.

Exam tip
Remember, you only need to learn one of these case studies on migration.

Check your understanding
For your chosen retirement case study, make a list of the specific facts that you need to learn.

Go online for answers

Exam practice
For a named example, explain the consequences of retirement migration on the destination. (6 marks)

Answers online

Exam tips
Remember to learn specific points (facts and figures) to use in your case study response.

Look back at the mark schemes for case study questions on page viii of the introduction.

Online

Chapter 16 A Tourist's World

Growth of the tourist industry

Section B – Remember that you have only been taught one of these topics. If you have been taught A Moving World, turn to page 107.

Leisure breaks

Revised

Tourism is measured by:

- visitor numbers
- the amount of value that the different types of tourism bring to an economy
- entry and exit surveys carried out at airports and ports.

Leisure breaks can be broken down into:

Beach holidays – the main purpose of the holiday is to spend time relaxing on a beach. This could be for sunbathing or for water sports. An international beach holiday would be a family from the UK going to Spain.

Activity holidays – these holidays provide travellers with a different challenge depending upon the type of activity that is involved. A local activity break could be a person from Ipswich attending a course on painting in Norfolk which would be a relaxing holiday.

Short city breaks – people travel to urban areas to enjoy a heritage and cultural break. A national city break would be a couple travelling from Birmingham to Liverpool for the weekend.

Health tourism – this type of tourism involves going on a trip which includes a holiday and some sort of health treatment.

Heritage and cultural tourism – people travel to visit sites of national and world importance to learn about the past. They may be sites of cultural importance such as the Hindu temples in Sri Lanka, or historical importance such as Windsor Castle.

Key terms

Tourism – the movement of people away from their main place of residence; it usually includes an overnight stay. It can be for leisure or business purposes or to visit family and friends.

Local tourists – people who travel in the area close to where they live.

National tourists – people who travel in their own country.

International tourists – people who travel between countries.

Business trips – people travelling to attend conferences and meetings.

Leisure trips – people travelling for enjoyment.

Seasonal tourism – tourism that can only take place at certain times of the year in a particular location, for example, beach holidays in the UK.

Exam tip

Learn the word **BASHH** as a way of remembering the different types of leisure breaks.

Exam tip

Learn an example of each type of leisure break.

Exam practice

Define the terms 'beach holiday' and 'short city breaks'.   (4 marks)

Answers online

Online

Resort development

The factors that have caused a growth in tourism

The growth of global tourism has been caused by a number of factors. The factors can be social, economic and political.

Key terms

Social factor – related to people.

Economic factor – related to money.

Political factor – related to the government.

Ageing population – a lot of people in the country are over the age of 50.

Package holidays – holidays where the price includes the transport and the accommodation.

All-inclusive – all transport, accommodation, food and drink is included in the price of the holiday.

Disposable income – the amount of money left out of a person's income to spend on luxury goods such as holidays after all necessities have been paid for.

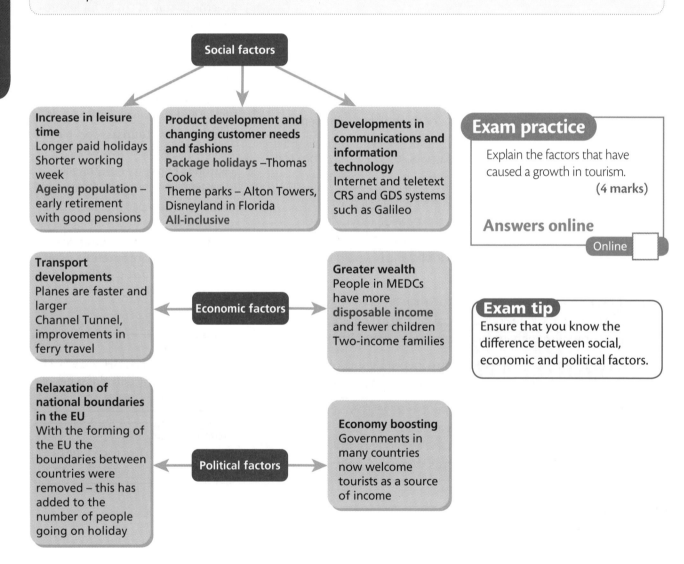

Social factors

Increase in leisure time
Longer paid holidays
Shorter working week
Ageing population – early retirement with good pensions

Product development and changing customer needs and fashions
Package holidays –Thomas Cook
Theme parks – Alton Towers, Disneyland in Florida
All-inclusive

Developments in communications and information technology
Internet and teletext
CRS and GDS systems such as Galileo

Transport developments
Planes are faster and larger
Channel Tunnel, improvements in ferry travel

Economic factors

Greater wealth
People in MEDCs have more disposable income and fewer children
Two-income families

Relaxation of national boundaries in the EU
With the forming of the EU the boundaries between countries were removed – this has added to the number of people going on holiday

Political factors

Economy boosting
Governments in many countries now welcome tourists as a source of income

Exam practice

Explain the factors that have caused a growth in tourism.
(4 marks)

Answers online

Online

Exam tip

Ensure that you know the difference between social, economic and political factors.

Tourist destinations offer a variety of physical and human attractions

Revised

Photograph A

Photograph B

Check your understanding

Tested

Study photographs A and B. Compare their physical and human attractions.

Go online for answers

Online

Key terms

Physical attractions – the factors about the natural landscape that appeal to people including guaranteed sunshine, white sandy beaches and warm sea if they were looking for a beach holiday.

Human attractions – the factors about the built environment that appeal to people including theme parks, ancient monuments and museums.

Check your understanding

Tested

Study the two photographs above. Compare the physical and human attractions of the two areas.

Go online for answers

Online

The Butler model of resort development

In 1980 R.W. Butler developed a model for resort development. The model has seven stages which he believes resorts go through as they become tourist destinations.

The table below shows how the Butler model can be applied to Blackpool, a resort on the north-west coast of the UK.

Stage of model	Definition	Date from	Event or fact
Exploration	• A small number of tourists visit a destination. • There are no impacts on the area. • There are physical and cultural attractions.	1735	Blackpool's first guest house opened, owned by Edward Whiteside. The only visitors were the landed gentry (rich people) who would ride on the beach and bathe in the sea.
Involvement	• Visitor numbers start to increase. • Hotels are built. • Transport is improved – railway lines built to the resort.	1819	Henry Banks opened the Lane's End Hotel which was Blackpool's first hotel. In 1846, the railway line was completed to Blackpool.
Development	• Visitor numbers continue to increase. • There are still mainly physical and cultural attractions. • Some human attractions are being built. • The host community are more involved with tourism.	1870	Central Pier opened with open-air dancing for everyone. New promenade was opened to the south which linked the different areas of Blackpool together.
Consolidation	• The number of tourists continues to increase but not as quickly. • Transport routes and access to the resort have been improved. • Most local people now work in the tourist industry; local economy relies on tourism income. • There are many facilities for tourists which are beginning to impact on the environment.	1912	Many attractions were built such as the Grand Theatre, Church Street. Blackpool illuminations were first switched on in 1912.
Stagnation	• The facilities and services become old and run down. • The negative impacts on the environment are becoming more obvious. • Visitor numbers are declining. • The host community begins to resent the tourists.	1986	The Sandcastle (an indoor swimming pool) and Blackpool Zoo opened. But visitor numbers were starting to decline.
Decline	• Tourist numbers start to decline dramatically. • Many people lose their jobs.	1987	Annual day visits declined from 7.4 million to 3.9 milliion.
Rejuvenation	• This usually involves investing a lot of money to improve facilities and amenities. • The resorts have to be made up to date.	2004	Around 11,000,000 people visited Blackpool.

Exam practice

Explain the development of Blackpool in terms of the Butler model of resort development. **(4 marks)**

Answers online

Impacts of tourist industry growth

Tourism is having an impact (effect) on popular tourist destinations in countries at different levels of development. These effects can be either positive or negative.

Key terms

Social impacts – impacts on the lives of the people who live in the area.

Economic impacts – impacts on the economy of the area.

Environmental impacts – impacts on the landscape of the area.

Direct impacts – impacts that have an immediate effect.

Indirect impacts – impacts that cause changes over time.

The social impacts (effects) of tourism
Revised

Area	Social impacts of tourism
Zanzibar, an island off the east coast of Africa, a Muslim country	Tourists are mainly from western countries which have different moral codes and culture to the local inhabitants, for example, they do not cover their shoulders. This causes offence to the local people.
Dubai – an Arab country in the Middle East	Many western business people now travel to Dubai for conferences. Men and women are expected to be fully clothed at all times and no form of contact should be made in public. These cultural differences have caused some issues for the business people who have to maintain the moral code of Dubai.
Malham – a small village in the Yorkshire Dales	In Malham, many of the houses are second homes. Second homes are only lived in for part of the year and the people who own them often do not shop in the local shop or use other local facilities such as the school and church. Malham has a more frequent bus service to the local town of Skipton in the summer months because the tourists use it.
The Maldives – a group of islands in the Indian Ocean	The Maldives have been developed as enclave resorts. This means that the tourists only mix with the local people who are employed at the resorts. The majority of the population live on the 'home islands' where the tourists are not allowed. Tourist money has also been used to improve education and services for the local people.

The economic impacts (effects) of tourism
Revised

Area	Economic impacts of tourism
Machu Picchu in Peru	This destination generates $40 million a year in income for the Peruvian government from the tourists who visit the site.
Zanzibar – an island off the east coast of Africa, a Muslim country	In Zanzibar, $220 million a year comes into the country from tourism. This has a positive effect on the GDP of the country.
Cyprus – an island in the Mediterranean	20 per cent of the GDP is provided by tourist income. This is the money that tourists spend on accommodation and food and drink when they are in the country. The **multiplier effect** means even more money is generated in the country.

Key terms

The multiplier effect – when a dollar injected into the tourist destination's economy circulates through that economy several times. For example, $500 spent at a hotel is then re-spent by the hotel to purchase food from the local farmers. The farmers use the money they are paid by the hotel to buy their supplies.

Leakage – money lost from the economy. For example, t-shirts bought from another country are sold within the country. The money that was spent on buying the t-shirts has leaked from the country's economy.

The environmental impacts (effects) of tourism

Area	Environmental impacts of tourism
Cyprus – an island in the Mediterranean	Due to the development of tourism in certain parts of Cyprus, Lara Beach on the Akamas Peninsula has been protected for Green and Hawksbill Turtles to lay their eggs.
Zanzibar – an island off the east coast of Africa, a Muslim country	Stone Town, in Zanzibar, has being designated as a World Heritage Site. This is as a direct result of the number of tourists who were visiting it. This has given it protection from tourism development. Development can take place only if it is to help to conserve the buildings in the town. Money from tourism has also been used to improve the buildings of Stone Town, for example, the Zanzibar Serena Hotel.
The Maldives – a group of islands in the Indian Ocean	The enclave resorts on the Maldives have caused problems for the coral reefs that surround them. There has been a decrease in fish numbers and species. This is due to the environmental problems caused by tourism due to sewage disposal and the tourists themselves swimming close to the coral reef. When new resorts are developed, the government insists that the sewage is dealt with in an eco-friendly way and that tourists are made aware of the environmental damage that they cause.
Lulworth Cove, Dorset	The environmental impacts of tourism at Lulworth Cove in Dorset have been minimised by the use of tourist money. People pay to park their cars at Lulworth Cove. This money has been used to build a Heritage centre which teaches the tourists about the area. It has also been used to improve the signage and to re-lay the footpaths which had become worn due to the sheer numbers of tourists who use them.

Exam tip

You could be asked questions that require recall of knowledge. You should learn at least one specific social, economic and environmental impact, both positive and negative.

Check your understanding

Tested

1 Define the terms social, economic and environmental impacts.

2 Draw three spider diagrams to show the impacts of tourism: one for social impacts, one for economic impacts and one for environmental impacts. Include at least one positive and negative impact in each one.

Go online for answers

Online

Exam practice

Explain the impacts of tourism on the environment. Use examples in your answer. **(4 marks)**

Answers online

Online

Exam tip

You may be asked questions about the effects of tourism on a photograph of an area you do not know. In this case you will have to apply your knowledge about general effects to the area shown.

Hotel worker in the Maldives
I work long hours at the resort and rarely see my family who live on the 'home islands'. I only wish that the pay was better.

Malham shop owner
I can open the shop all week in the summer because of the numbers of tourists in the village. This is great for me and the local community.

Attitudes to tourism vary between individuals, organisations and governments

Lulworth Heritage Centre manager
People who come to the area usually start their visit here. The centre teaches them about the area – educated tourists cause less damage.

Peruvian government official
Tourism is great for the country as it brings in lots of money.

Photograph A

Photograph B

Exam practice

Study the photographs A and B.
Photograph A is of a ski resort in the French Alps.

1 Describe the physical and human attractions of the area.

2 Look again at the photograph.

  a What type of leisure breaks could take place in this area?

  b Give reasons for your answer to part (a).

3 Explain the effects of tourism on this area.

Photograph B is of a tourist area in the UK.

4 Describe the type of leisure tourism that could take place in this area.

5 Explain the effects of tourism on this area.

Answers online

Online

Ecotourism

Ecotourism

'Footsteps' is an **ecotourism** enclave resort in The Gambia which is located close to the village of Gunjur. It is open all year round and many of the tourists come from the UK.

The green boxes describe how it protects the environment; the purple boxes describe how it benefits the local community.

Footsteps has gardens where it grows vegetable crops. These gardens provide all the food for the tourists. There are also fruit trees (banana, mango, orange) which grow around the huts.

Ducks are kept which provide eggs.

Electricity is produced using the wind and the Sun. The solar-powered freezer has reduced the use of propane gas.

The water for the swimming pool is filtered through reed beds to get rid of impurities instead of using chlorine in the water.

The huts are made from local wood. The furniture is made by craftsmen from Gunjur from local wood.

Footsteps gives 20 per cent of its profit to the local community.

There is a shop at Footsteps which sells Batik work which is made at Footsteps by local women.

Key terms

Ecotourism – responsible tourism that conserves the environment and provides benefits to the local people.

Sustainable tourism – tourism which leads to the management of all resources in a way that meets the needs of present tourists and local people while protecting the needs of future generations.

Footsteps, Gunjur, The Gambia

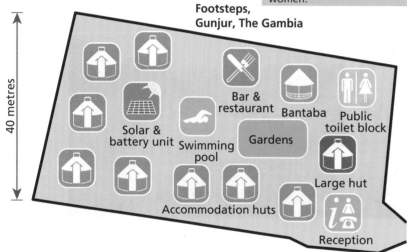

40 metres

80 metres

Solar & battery unit

Bar & restaurant

Bantaba

Public toilet block

Gardens

Swimming pool

Large hut

Accommodation huts

Reception

Check your understanding

1 State three ways in which Footsteps helps improve the lives of local people.

2 State three ways in which Footsteps helps to protect the environment.

Go online for answers

The toilets at Footsteps are composting toilets. This means that after harmful substances have been removed, the waste can be used as compost on the garden.

Water is very scarce. It comes from wells and is stored in water tanks. Solar-powered pumps are used to fill up the water tanks. The water that has been used by guests for washing is known as grey water. It is filtered and then used to irrigate the fruit and vegetables grown in the gardens.

All 22 of the staff are from the local village of Gunjur. The employees receive training and are paid for the whole of the year even if the hotel does not have any guests. Most locally employed staff are only paid in the season when the hotel is busy. The staff also get medical and dental care.

↑ **Figure 1 Footsteps, Gunjur, The Gambia**

Exam practice

For a named ecotourism destination explain the reasons why it is good for the environment and the local community. **(6 marks)**

Answers online

Online